The Competitive Status of the U.S. Pharmaceutical Industry

The Influences of Technology in Determining International Industrial Competitive Advantage

Prepared by the Pharmaceutical Panel,
Committee on Technology and
International Economic and Trade Issues

Office of the Foreign Secretary,
National Academy of Engineering

Commission on Engineering and
Technical Systems,
National Research Council

Charles C. Edwards, Chairman
Lacy Glenn Thomas, Rapporteur

NATIONAL ACADEMY PRESS
Washington, DC 1983

National Academy Press • 2101 Constitution Avenue, N.W • Washington, DC 20418

This project was supported under Master Agreement No. 79-02702, between the National Science Foundation and the National Academy of Sciences.

Library of Congress Catalog Card Number 83-50568

International Standard Book Number 0-309-03396-9

First Printing, August 1983
Second Printing, August 1984

Panel

CHARLES C. EDWARDS (Chairman), President, Scripps Clinic
and Research Foundation
KENT BLAIR, Vice-President, Donaldson Lufkin & Jenrette
WILLIAM NEILL HUBBARD, JR., President, Upjohn Company
PETER BARTON HUTT, Partner, Covington and Burling
PHILIP RANDOLPH LEE, Professor of Social Medicine,
University of California Medical School, San Francisco
ARTHUR M. SACKLER, Research Professor, New York Medical
College, Publisher, Medical Tribune Newspapers
LEWIS HASTINGS SARETT, Senior Vice-President, Merck & Co.,
Inc. (retired)
WILLIAM MICHAEL WARDELL, Professor, Department of
Pharmacology, University of Rochester Medical Center
PAUL F. WEHRLE, Professor of Pediatrics, University of
Southern California
ALBERT P. WILLIAMS, Director, Health Science Program, The
Rand Corporation
RICHARD WOOD, Chairman of the Board and Chief Executive
Officer, Eli Lilly & Company
ALEJANDRO ZAFFARONI, President and Director of Research,
ALZA Corporation

Rapporteur

LACY GLENN THOMAS, Professor, Graduate School of Business,
Columbia University

iii

Additional Participants

JAMES ANDRESS, Vice President, Corporate Planning, Abbott
  Laboratories
J. RICHARD CROUT, Director, Bureau of Drugs, Food and Drug
  Administration
ELI FROMM, House Science, Research and Technology
  Subcommittee of the Science and Technology Committee, U.S.
  Congress
LEO R. McINTYRE, Office of Basic Industries, U.S. Department
  of Commerce
PAUL MEYER, Assistant Director of Public Affairs for Public
  Policy, Pfizer, Inc.
DUFFY MILLER, Editor, PMA Newsletter, Pharmaceutical
  Manufacturers Assoc.
SUMIYE OKUBO, Policy Analyst, Division of Policy Research and
  Analysis Scientific, Technological, and International Affairs,
  National Science Foundation
ROLF PIEKARZ, Senior Policy Analyst, Division of Policy
  Research and Analysis, Scientific, Technological, and
  International Affairs, National Science Foundation
ALAN RAPOPORT, Policy Analyst, Division of Policy Research
  and Analysis, Scientific, Technological, and International
  Affairs, National Science Foundation
C. MELVIN STONE, Director, International Economic Research,
  Pharmaceutical Manufacturing Association
JULIUS SPIRO, Economist, U.S. Department of Labor

Consultant

BENGT-ARNE VEDIN, Research Program Director, Business and
  Social Research Institute, Stockholm, Sweden

Staff

HUGH H. MILLER, Executive Director, Committee on Technology
  and International Economic and Trade Issues
MARLENE R.B. BEAUDIN, Study Director, Committee on
  Technology and International Economic and Trade Issues
ELSIE IHNAT, Secretary, Committee on Technology and
  International Economic and Trade Issues
STEPHANIE ZIERVOGEL, Secretary, Committee on Technology
  and International Economic and Trade Issues

iv

E. RAY McCLURE, Program Leader, Precision Engineering Program, Lawrence Livermore Laboratory and Chairman, CTIETI Machine Tools Panel

BRUCE S. OLD, President, Bruce S. Old Associates, Inc. and Chairman, CTIETI Ferrous Metals Panel

MARKLEY ROBERTS, Economist, AFL-CIO

LOWELL W. STEELE, Consultant--Technology Planning and Management*

MONTE C. THRODAHL, Vice-President, Technology, Monsanto Company

---

*Formerly, Staff Executive, General Electric Company.

Preface

In August 1976 the Committee on Technology and International Economic and Trade Issues examined a number of technological issues and their relationship to the potential entrepreneurial vitality of the U.S. economy. The committee was concerned with:

* Technology and its effect on trade between the United States and other countries of the Organization for Economic Cooperation and Development (OECD);
* Relationships between technological innovation and U.S. productivity and competitiveness in world trade; impacts of technology and trade on U.S. levels of employment;
* Effects of technology transfer on the development of the less-developed countries (LDCs) and the impact of this transfer on U.S. trade with these nations; and
* Trade and technology exports in relation to U.S. national security.

In its 1978 report, Technology, Trade, and the U.S. Economy,* the committee concluded that the state of the nation's competitive position in world trade is a reflection of the health of the domestic economy. The committee stated that, as a consequence, the improvement of our position in international trade depends primarily upon improvement of the domestic economy. The committee further concluded that one of the major factors affecting the health of our domestic economy is the state of industrial innovation. Considerable evidence was presented during the study to indicate that the innovation process in the United States is not as vigorous as it once was. The committee recom-

---

*National Research Council, 1978. Technology, Trade, and the U.S. Economy. Report of a Workshop held at Woods Hole, Massachusetts, August 22-31, 1976. National Academy of Sciences, Washington, D.C.

mended that further work be undertaken to provide a more detailed examination of the U.S. government policies and practices that may bear on technological innovation.

The first phase of study based on the original recommendations resulted in a series of published monographs that addressed government policies in the following areas:

* The International Technology Transfer Process.*
* The Impact of Regulation on Industrial Innovation.*
* The Impact of Tax and Financial Regulatory Policies on Industrial Innovation.*
* Antitrust, Uncertainty, and Technological Innovation.*

This report on the pharmaceutical industry is one of six industry-specific studies that were conducted as the second phase of work by this committee. Panels were also set up by the committee to address automobiles, electronics, ferrous metals, machine tools, and fibers, textiles, and apparel. The objective of these studies was to (1) identify global shifts of industrial technological capacity on a sector-by-sector basis, (2) relate those shifts in international competitive industrial advantage to technological and other factors, and (3) assess future prospects for further technological change and industrial development.

As a part of the formal studies, each panel developed (1) a brief historical description of the industry, (2) an assessment of the dynamic changes that have been occurring and are anticipated as occurring in the next decade, and (3) a series of policy options and scenarios to describe alternative futures for the industry. The primary charge to the panel was to develop a series of policy options to be considered by both public and private policymakers.

The methodology of the studies included a series of panel meetings involving discussions between (1) experts named to the panel, (2) invited experts from outside the panel who attended as resource persons, and (3) government agency and congressional representatives presenting current governmental views and summaries of current deliberations and oversight efforts.

The drafting work on this report was done by Dr. Lacy Glenn Thomas, Columbia University. Professor Thomas was responsible for providing research and resource assistance as well as producing a series of drafts, based on the panel deliberations, that were reviewed and critiqued by the panel members at each of their three meetings.

----

*Available from the National Academy of Engineering, Office of the Foreign Secretary, 2101 Constitution Avenue, N.W., Washington, D.C. 20418.

Contents

The Competitive Status
of the U.S. Pharmaceutical
Industry

Summary

The U.S. pharmaceutical industry has for decades been one of the most profitable and rapidly growing sectors of the American economy. Its continuing expansions of output, productivity, and jobs have been achieved alongside price increases that have been more moderate than the general rate of inflation. Together with other high-technology industries, it has played an important role in generating exports and net trade surpluses. Additionally, new pharmaceuticals have made significant contributions to improved health and to the control of escalating medical costs.

On the basis of these achievements, the pharmaceutical industry has maintained an image of immunity from the deterioration of competitive position besetting many sectors of the American economy, such as automobiles, steel, textiles, and consumer electronics. Unfortunately, this image is apparently exaggerated, and probably false. Data compiled by this study indicate a clear relative deterioration in the foundation of pharmaceutical competitive position--the research efforts necessary for discovery and introduction of new patented drugs.

Persistence of the image of unchallenged American preeminence in pharmaceuticals would appear to be based on two rather unique features of the industry. In the first place, the time lapse between strategic decisions by ethical drug firms and the impact of these decisions on the market is particularly long. The most critical of these decisions involves investments in discovery of new patentable drugs, yet basic pharmaceutical research performed today may well not produce marketed products within this century, and even drugs now being synthesized will on average not be introduced into the United States until the mid-1990s. As a consequence, deteriorations now occurring in the relative innovative abilities of American pharmaceutical firms will not be visible in product markets for several years and not fully felt for as long as two decades.

1

A second factor masking the relative decline of U.S. firms is the inherent potential for growth in the pharmaceutical industry. This is a time of major advances in the basic sciences of human health, developments that have opened up significant possibilities for new drug products and associated sales growth. Given this progress in basic sciences, U.S. pharmaceutical firms will almost inevitably be, and in fact are, innovative, growing, and profitable. The performance of the U.S. ethical drug industry is thus quite different from those less fortunate sectors of the U.S. economy that have been damaged by escalating energy prices or that are threatened by low-wage competition from developing nations. Comparisons among different sectors of the U.S. economy, however, are not useful for evaluation of the performance either of corporate management or national industrial policy, precisely because the underlying potential for growth or decline in a particular industry is in large measure uncontrollable. Far more relevant comparisons may be made among U.S. and foreign firms within the same industry, all of which have access to similar technological opportunities. From this perspective of relative U.S.-foreign competitive performance, a declining U.S. share of a growing industry is as much a concern for U.S. industrial policy as a declining share of an industry undergoing retrenchment.

Not every reduction in the U.S. market share of an industry, of course, is indicative of managerial or public policy failures. For example, the almost economy-wide loss of U.S. shares of major markets to Europe during the 1950s represented the recovery of those war-damaged economies to normal levels of output rather than any faltering of American economic achievement. Likewise, the gradual diffusion overseas of production by those industries that extensively use unskilled labor has been interpreted as the basic consequence of free trade and represents efficient reallocation of resources. This report does not address the question of an appropriate U.S. share of the world pharmaceutical industry. Nonetheless, several circumstances--the traditional importance of U.S. firms within this industry, the excellence of American research in basic biomedical sciences, the enormous expenditures of the U.S. government to fund this basic research, and the general importance of high-technology industries for the U.S. trade balance--all suggest that the relative decline in U.S. pharmaceutical competitive position should be cause for further inquiry, if not concern.

FINDINGS ON THE U.S. COMPETITIVE POSITION

Because 14 or more years can separate the synthesis of new ethical drugs and the ultimate marketing of these substances, available data at various stages in the pharmaceutical development process allow the examination of the expected evolution of competitive position over the near-term future. The foundation of national pharmaceutical competitive advantage lies in successful innovation. Thus current research efforts provide a forecast of future sales, earnings, and jobs in the industry. By examining different segments of the innovation process--from R&D expenditure, to drugs entering clinical trials, to marketed drugs, to sales and market structure for new drugs--the existing and expected future competitive patterns may be simultaneously compared.

The basic findings on the U.S. position at various stages of the pharmaceutical innovation process are summarized below and discussed in detail in the second chapter of the report.

• The U.S. share of world pharmaceutical R&D expenditures has fallen from greater than 60 percent during the 1950s to less than 30 percent in 1982.
• The number of new drugs entering U.S. clinical trials and owned by U.S. firms has steadily dropped from a yearly average of 60 in the mid-1960s to about 25 per year in 1982. In contrast, the number of foreign-owned drugs undergoing comparable trials has remained almost constant at about 20 per year.
• The U.S.-owned share of new drug introductions has remained roughly stable in most major markets, with generally a 60 percent share of U.S. introductions and a 22 percent share of worldwide introductions.
• The percentage of world pharmaceutical production occurring in the United States has fallen from 50 percent in 1962, to 38 percent in 1968, to 27 percent in 1978.
• The share of pharmaceutical sales by U.S.-owned firms fell slightly in major markets during the 1960s and has been roughly constant since. In our domestic market, the share of U.S. firms was 87 percent in 1965 and 80 percent in 1982.
• Smaller U.S. pharmaceutical firms self-originate fewer new drugs than before 1960 and are increasingly dependent on foreign firms for licensed new products, though licensed products still account for less than half of drug introductions by small firms.
• During the 1970s, European pharmaceutical firms established a broad multinational base in the U.S. domestic market that will in the near future be used for direct marketing of European pharmaceutical innovation.
• The U.S. share of world pharmaceutical exports has fallen from greater than 30 percent before 1960 to less than 15 percent today.

An overview of these trends indicates a marked drop for the U.S. presence in world pharmaceutical markets around 1960, followed by stability in the U.S. share of new drug introductions and sales (outputs of the innovative process). In contrast, the U.S. share of R&D expenditures and drugs entering clinical trials (inputs of the innovative process) has continued to decline, strongly suggesting an eventual further decline in U.S. shares of introductions, sales, and exports.

DETERMINANTS OF NATIONAL PHARMACEUTICAL COMPETITIVE ADVANTAGE

Sources of these trends in competitive postion can be segregated into two categories--those factors that generally affect the entire U.S. economy and those factors that have unique impact on the pharmaceutical industry. Numerous studies have documented an almost economy-wide deterioration in competitive position for American firms against their foreign counterparts. As is discussed in the third chapter of this report, many of the declines in U.S. pharmaceutical competitive position listed above can be attributed to whatever factors have led to the relatively poorer performance of the U.S. economy in the aggregate.

Two aspects of pharmaceutical competitive position, however, are atypical from the general U.S. industrial experience. The first unique feature has been the precipitous drop in the proportion of world drug production located within U.S. boundaries, a decline wholly unmatched in other segments of the chemical industry. Foreign non-tariff trade barriers such as discriminatory safety regulations and pricings by public health authorities are apparently the predominant cause of this divergent trend. Second and even more significant for the future economic strength of U.S. ethical drug firms, the steady decline in the American share of world pharmaceutical R&D efforts is markedly more severe than comparable changes in world R&D shares for other U.S. industries. Again, factors specific to the ethical drug industry should be invoked to explain this distinctive performance. Factors that have contributed to this important trend are (in no particular order):

- Foreign non-tariff trade barriers, mentioned above.
- U.S. Food and Drug Administration (FDA) regulations, imposing significant costs and delays on the research efforts of U.S. firms.
- Patent laws, differing among developed nations in the extent of market protection provided to innovators. In this regard, U.S. patent policy is more restrictive than that of certain nations, but more generous than that of some others.

•   Liability regimes for consumer product claims that are more cumbersome and risky in the United States than in certain competitor nations.
•   Antitrust policies that may prevent attainment of economies of scale in pharmaceutical research through mergers among U.S. firms.
•   Tax incentives for conduct of research.

Determination of the specific relative importance of each factor is beyond the scope of this study. The basic conclusion to be drawn, however, remains that the overall balance of several (possibly conflicting) government policies provides a relatively more favorable environment abroad for pharmaceutical research. Careful evaluation of these and other government policies with the goal of encouraging innovation is needed.

OPTIONS FOR AMERICAN POLICY

The study identifies a variety of policy options to counteract the causes of decline in the competitive position of the United States pharmaceutical firms.

Trade Options

FDA policy prohibiting the export of unapproved new drugs, and thus requiring United States companies to manufacture these products abroad, should be revised by regulation to permit the export of pharmaceutical chemicals for such use. The prevalence of foreign trade barriers that favor domestic products over American drugs should be investigated to determine an appropriate United States response.

Domestic Economic Options

Legislation should be enacted to restore the amount of patent time lost as a result of FDA regulatory requirements. Antitrust policy should be reconsidered to determine whether it discourages mergers that would make U.S. pharmaceutical companies more effective competitors on a worldwide scale. Research tax credits should be expanded to include research-related expenditures not now eligible for the investment tax credit. Research and development expenditures incurred in the United States should be allocated solely to the United States income of the taxpayer. The impact of product liability on the pharmaceutical industry should be studied in an attempt to reduce this disincentive to research.

Regulatory Options

The recommendations of the recent report of the Commission on the Federal Drug Approval Process should be implemented by FDA through administrative changes in current regulations as rapidly as possible. Adoption of these recommendations would expedite the IND and NDA review process, thus reducing the size of the investment needed to develop new pharmaceutical products and increasing the return on such investment. Improvements in the drug approval process can be made without any reduction in public health protection and can be expected to result in more rapid availability of important new drugs to combat serious diseases for which effective drugs are not currently available.

Overview
of U.S. Pharmaceutical Industry

The importance of research and innovation for competition among major pharmaceutical firms places the ethical drug industry in a select grouping of high-technology industries. The most distinctive feature of pharmaceutical innovation lies in the spending strategies of the major firms—high rates of investment in R&D expenditures (as percentages of sales and profits), relatively high rates of spending for basic research, and little government financing of industrial R&D. These trends are illustrated in Table 1-1 and indicate that, while one or more of these features are present in other industries, rarely are all three. The pharmaceutical industry, along with the computer, photographic, and specialized machinery industries, all spend more than 50 percent of their recorded profits on research and development.

On the basis of this innovation, American firms were predominant in world markets during the period 1950 to 1960, accounting for a large majority of research expenditures and new products, over half of world pharmaceutical production, and one third of international trade in medicinals. American preeminence persisted, though in attenuated degree, through the 1960s. In the past decade, however, the competitive advantage of American firms has been not only reduced, but apparently eliminated. This study seeks to define and document these changes of competitive position within the multinational pharmaceutical industry, to determine why these changes have occurred, and to suggest an array of policy options to address the relative decline. This first chapter provides a primer on competition within the ethical drug industry.

EMERGENCE OF THE MODERN
PHARMACEUTICAL INDUSTRY

The drug industry before 1930 was profoundly different from that of today. Innovation was infrequent and externally derived, and

TABLE 1-1   Research Attributes of Various U.S. Based Industries, 1977

Industry	Basic Research as Percentage of Total R&D	R&D as Percentage of Sales	Government Funding as Percentage of R&D Funds
Drugs and medicines	11.4	6.2	1.0
Industrial chemicals	9.7	3.6	19.0
Food and kindred products	5.2	0.4	na
Stone, clay, and glass products	14.0	1.2	na
"Other" chemicals	9.6	2.1	na
Petroleum refining and extraction	5.3	0.7	8.1
Communications equipment	5.2	7.6	43.1

SOURCE: *Research and Development in Industry, 1977.* Washington, DC, National Science Foundation, 1979.

firms manufactured a limited number of unpatented products which were largely marketed without prescription directly to consumers. The mix of products available to consumers has been described by a pharmaceutical executive, Henry Gadsden of Merck, when he described the nature of the market in the 1930s:

> You could count the basic medicines on the fingers of your two hands. Morphine, quinine, digitalis, insulin, codeine, aspirin, arsenicals, nitroglycerin, mercurials, and a few biologicals. Our own Sharp and Dohme catalog did not carry a single exclusive prescription medicine. We had a broad range of fluids, ointments, and extracts, as did other firms, but we placed heavy emphasis on biological medicines as well. Most of our products were sold without a prescription. And 43 percent of the prescription medicines were compounded by the pharmacist, as compared with 1.2 percent today.[1]

None of these products mentioned by Gadsden had resulted from research efforts of the pharmaceutical industry. Only a handful of drug discoveries from any source had been made by 1930 (principally salversan in 1908 for treatment of syphillis and insulin in 1922 for treatment of diabetes) and these discoveries were infrequent, unrelated, and unanticipated, and resulted from prolonged and tedious research. Nothing about these discoveries suggested a method of research or a mechanism of disease prevention that could be economically exploited for development of new pharmacological agents.

This non-innovative technological environment changed rapidly just before and during World War II, in a "therapeutic revolution" that transformed the industry. First, during the period 1930 to 1950, a series of natural products, particularly the vitamins and

hormones, were discovered, developed, and commercialized.[2]
These discoveries led to the conquest of scurvy, pernicious
anemia, beri-beri, and pellagra as well as significant endocrine
therapies. Second, the foundation was laid for modern research in
anti-infectives. The discovery of the therapeutic properties of
sulfanilamide by I. G. Farbenindustrie in 1935 and of penicillin by
Oxford scientists in 1940 indicated the possibilities for <u>systematic</u>
research in finding new sulfa drugs and new antibiotics. Neither
sulfanilamide nor penicillin were patentable at the time, having
been known discoveries with belated demonstration of therapeutic
properties. Nonetheless, the tremendous demand for anti-
infective agents by allied military forces during wartime made the
manufacture of these scarce substances a national priority. The
U.S. government spent almost $3 million to subsidize wartime
penicillin research and encouraged private construction of penicillin manufacturing plants by allowing accelerated depreciation.
The returns from sales of these and other drugs were subject to
wartime "excess profits" taxes, but at the conclusion of World War
II, federal penicillin plants were sold to private firms at half cost.

The simultaneous demonstration of new technological
opportunities and of potential profits combined to dramatically
change the pharmaceutical industry. The final step necessary for
the emergence of the industry in its modern form was a legal
mechanism to allow commercial exploitation of the new technological opportunities for biological products. This step occurred
with the 1948 decision of the U.S. Patent Office to grant a patent
for streptomyicin. A patent, of course, is a legal monopoly for 17
years over commercial exploitation of a new discovery. During
the period before expiration of the patent, the innovative firm
may charge prices above manufacturing costs and thus recoup
earlier research expenditures that led to the innovation. Rapidly,
a new form of competition emerged in the pharmaceutical
industry--competition through product development.

At the outset of the 1950s, pharmaceutical competition
remained largely national in scope, with the significant exception
of the Swiss multinationals. Economic linkages among the various
national pharmaceutical industries were largely confined to international trade, and even then were relatively unimportant.
Imports amounted to less than 10 percent of domestic consumption in the major industrial nations, again with the exception of
Switzerland. Firms engaged in new product development faced
essentially three methods for foreign distribution of their
innovations:

• Exports--domestic production by the innovating firm for sale
abroad through local distributors.

TABLE 1-2 Domestic and Foreign Sales of U.S. Owned
Pharmaceutical Firms, Various Years (percentages)

Year	Domestic	Foreign
1956	88	12
1961	73	27
1966	71	29
1971	66	34
1976	60	40
1978	57	43

NOTE: Table statistics are based on sales of human dosage. They
exclude sales of bulk drugs and veterinary drugs.

SOURCE: *Annual Survey Reports* (Washington, DC: Pharmaceutical Manufacturers' Association, various years).

• Licensing--production abroad by a foreign firm with profits
shared between the innovating firm and the producer.
• Multinational expansion--production abroad by a subsidiary
of the innovating firm.

Starting in the 1950s, American firms began and Swiss firms
continued substantial multinational expansion of operations (for
data on U.S. firms, see Table 1-2). The presence of tariff and
regulatory barriers imposed by foreign governments, greater
physician and consumer acceptance of local production sources,
and a general tendency toward vertical integration by pharmaceutical firms made reliance on exports a less viable and
profitable strategy. In general, the choice between licensing and
multinational investment depended on the breadth of a firm's
product line. American and Swiss firms that enjoyed a surge in
the number of new patented drugs during the 1950s and 1960s were
able to spread the substantial overhead costs of direct foreign
investment over the several drugs distributed abroad by their
firms, making direct investment relatively less burdensome.
Non-Swiss, European, and Japanese firms with narrower product
lines that might have attempted direct investment abroad would
have been forced to cover these overheads entirely from sales of
just a few drugs--a potentially unprofitable endeavor. An
additional factor that limited non-Swiss, European, and Japanese
direct investment arose from the economic devastation of World
War II and the financial burdens of reconstruction. The resulting
pattern of multinational expansions can be seen in Table 1-3.
After 1960 the costs of developing commercially viable new
drugs dramatically increased. One consequence of this important
trend has been that larger earnings, available only from a larger
market, were essential to cover the greater costs of R&D for each

Table 1-3 Multinational Structure of Major Pharmaceutical Markets, 1973

Nationality (Ownership)	Nationality (location)									
	USA	Japan	Germany	France	Italy	Spain	UK	Brazil	Mexico	Canada
USA	*	–	12.6	17.4	15.8	14.4	38.4	35.4	49.6	63.4
Japan	–	*	–	–	–	–	–	–	0.1	–
West Germany	1.0	4.6	*	4.5	7.6	10.4	7.1	13.3	7.4	2.0
France	–	0.3	1.9	*	3.7	3.1	4.6	3.4	3.5	2.2
Italy	–	–	0.2	0.1	*	2.7	0.1	4.6	2.7	–
Switzerland	12.6	3.3	9.3	9.2	10.9	8.9	10.7	10.3	9.4	11.1
UK	2.2	2.3	1.8	3.5	5.1	1.2	*	1.9	3.5	4.9
Netherlands	0.1	0.4	1.8	1.2	0.3	0.8	1.7	1.1	1.3	–
Sweden	0.1	0.2	0.4	0.3	–	–	0.7	0.3	0.2	0.3
Other Foreign	–	0.1	1.7	1.6	1.1	2.1	0.4	–	–	0.8
Total Foreign	16.0	23.4	29.7	37.8	44.5	43.6	63.7	70.3	77.7	84.7
Local Ownership	84.0	76.6	70.3	62.2	55.5	65.4	36.3	29.7	22.3	15.3
Total	100.0%	100.0%	100.0%	100.0%	100.0%	100.0%	100.0%	100.0%	100.0%	100.0%

NOTE: Asterisk takes place of local percentages in top half of table. Local percentages are given separately in the bottom half as Local Ownership.

SOURCE: Barrie Jones, *The Future of the Multinational Pharmaceutical Industry to 1990.* New York: John Wiley, 1977.

compound. This industrial need to cover rising research costs, along with the almost universal cross-cultural use of pharmaceuticals, and the dramatic expansion of third-party payments for health-care costs combined to insure the emergence of a world market in ethical drugs. While this world market is severely fragmented due to non-tariff barriers to trade and due to differing national regulations, it is nonetheless increasingly inescapable that the competitive vitality of the major pharmaceutical firms depends on distribution of new products on a worldwide scale.

NATURE OF PHARMACEUTICAL COMPETITION

Prior to the therapeutic revolution of the 1940s, the pharmaceutical industry exhibited three distinct divisions, each with its own form of competition. The first subindustry, proprietary drugs, or over-the-counter (OTC) medicines as they are also called, encompasses products sold directly to consumers without prescription in the context of extensive advertising. Competition in this segment of the pharmaceutical industry depends largely on marketing of established brands with occasional new product development. New proprietary drugs rarely represent breakthroughs in treatment and often are simple reformulations of existing therapies that facilitate consumer convenience or are products switched from prescription to OTC status as a result of the U.S. FDA OTC drug review.[3] Proprietary drugs are thus characterized by high advertising intensity but a very low research intensity. Sales of proprietary drugs have grown at a markedly slower rate than other pharmaceutical sales and currently comprise less than 15 percent of total industry sales, as can be seen in Table 1-4.[4] About 550 firms in the U.S. produce and distribute exclusively OTC medicines.[5]

The second division of the industry, generic products or multisource drugs, exhibits the classical form of market competition. Generic drug products are off-patent, well-established compounds that are produced as standardized commodities by more than one firm. Generic products are generally unadvertised and usually subject to price competition among the various producers with the result of low profit margins for generic producers. Multisource drugs accounted for about 45 percent of ethical drug sales within the United States in 1979, though only 7 percent of these sales (or 3 percent of all drug sales) were achieved by the smaller, non-research-intensive firms. About 600 additional firms produce generic drugs in the United States. Almost all of these firms have exclusively domestic distribution, and many sell only to regional markets. Most generic drug houses have annual sales of less than $10 million.[6]

TABLE 1-4 Market Divisions of the Domestic U.S. Pharmaceutical Industry, Various Years (millions of dollars)

Year	Prescription Drugs	All Medicines	Prescription Drugs as a Percentage of all Medicines
1929	190	600	32
1949	940	1,640	57
1969	5,395	6,480	83

SOURCE: Peter Temin, *Taking Your Medicine: Drug Regulation in the United States.* Harvard University Press, Cambridge, 1980.

This study focuses on the remaining segment of the pharma-ceutical industry, <u>patented drugs,</u> distributed by prescription. Patented drugs represent the driving force of the modern pharma-ceutical industry and are responsible for the spectacular growth in sales since 1940. About 150 firms conduct research for and produce patented drugs in the United States. Only 20 of these firms have significant U.S.-based multinational operations, and about an equal number (20) are U.S.-located operations of foreign-owned multinational firms. The remaining firms have largely domestic sales, and some have very small research facilities. Industrial competition in this segment of the industry is quite distinctive and occurs through corporate development of new patented therapies.

Under patent protection, firms that introduce new products are able in principle to earn large returns on their innovations. There are, however, two constraints on the abilities of firms to generate earnings through innovation. The first is that it is generally technically possible for another firm to produce compounds of similar therapeutic action, though with different and hence also patentable molecular structure. The second constraint is, of course, that pharmaceutical innovation is a highly uncertain process that does not predictably yield therapeutically, let alone commercially, important products. Numerous firms have expended substantial funds for pharmaceutical R&D without development of a commercially successful product. Table 1-5 provides a tabu-lation of U.S. sales in 1972 of all new medicinal chemical compounds introduced into the U.S. market in the mid-1960s. While a very few products enjoyed substantial commercial success, the vast majority of products were relative commercial failures and did not contribute significantly to defraying R&D costs.

Given that the majority of contemporary pharmaceutical sales are comprised of generic products and patented drugs (both sold through prescription) and that profit margins in the generic products division of the industry are relatively low, it is clear that many industry profits are drawn from sales of patented drugs.

TABLE 1-5   New Chemical Entities (NCEs) Introduced in
U.S. 1962-1968 by 1972 U.S. Domestic Sales

Sales $000	Number of Drugs
0- 999	33
1,000- 1,999	14
2,000- 3,999	9
4,000- 5,999	5
6,000- 7,999	3
8,000- 9,999	1
10,000-14,999	4
15,000-19,000	2
20,000-29,999	2
30,000-39,999	2
40,000-49,999	2
50,000-59,999	0
60,000-99,999	1
100,000+	1
Total	70

SOURCE: David Schwartzman, *Innovation in the Pharmaceutical Industry*, Johns Hopkins University Press, Baltimore, 1976.

Finally, given that most pharmaceutical innovations are commercially not very successful, it is clear that modern pharmaceutical firms depend crucially for positive cash flow on a small handful of successful innovations, as is demonstrated for the United States in Table 1-6. Failure to produce new products continuously to replace those that lose market share to imitation or on which patents expire would ultimately be devastating to the financial health of a pharmaceutical company. In short, competitive advantage in sales of patented drugs, by far the most financially lucrative segment of the modern pharmaceutical industry, depends crucially on the ability of the firm to produce a slow but steady stream of commercially successful new products through industrial innovation.

BENEFITS AND RISKS OF TECHNICAL CHANGE

The rapid introduction of novel and complex products in any industry presents both social benefits and social costs. Because ethical drugs directly affect the health and lives of millions of consumers, the nonmarket implications of pharmaceutical innovation are especially pronounced.

As regards benefits, modern pharmaceutical products have substantially contributed to modern treatment of ill health. In this context, Victor Fuchs has observed:

TABLE 1-6 Proportion of Total Domestic U.S. Pharmaceutical Sales Provided by Three Best Selling Products, Selected Pharmaceutical Corporations, Selected Years (percentages)

	1970	1975	1979
Abbott	36	33	28
American Home Products			
Ayerst	64	74	84
Wyeth	37	44	43
Bristol-Meyers			
Bristol	69	46	28
Mead-Johnson	40	38	37
Burroughs Wellcome	na	56	51
Ciba	47	na	55
Lederle	48	31	32
Lilly	46	60	43
Merck	35	44	44
Pfizer	52	65	65
Robins	43	45	46
Roche	80	80	70
Schering	42	48	40
Searle	45	49	44
Smith Kline	44	42	66
Squibb	28	31	23
Upjohn	47	50	56
Warner-Lambert			
Warner	53	na	na
Parke-Davis	25	27	22

SOURCE: Merck & Co., Inc., MSD Strategic Planning and MSD Marketing and Sales Research, West Point, PA.

Original data from Intercontinental Medical Statistics (IMS), Inc., Ambler, PA.

Drugs are the key to modern medicine. Surgery, radiotherapy, and diagnostic tests are all important, but the ability of health care providers to alter health outcomes--Dr. Walsh McDermott's "decisive technology"--depends primarily on drugs. Six dollars are spent on hospitals and physicians for every dollar spent on drugs, but without drugs the effectiveness of hospitals and physicians would be enormously diminished.

Until this century the physician could with confidence give a smallpox vaccination, administer quinine for malaria, prescribe opium and morphine for the relief of pain and not much more. A quarter-century later the situation was not much different. Some advances had been made in surgery, but the death rates from tuberculosis, influenza and pneumonia, and other infectious diseases were still extremely high. With the introduction and wide use of sulfonamide and penicillin, however, the death rate in the United States from influenza and pneumonia fell by more than 8 percent annually from 1935 to 1950. (The annual rate of decline from 1900 to 1935 had been only 2 percent.) In the case of tuberculosis, while some progress had been made since the turn of the century, the rate of decline in the death rate accelerated appreciably after the adoption of penicillin, streptomycin, and PAS (paraamino-salicylic acid) in the late 1940s and of isoniazid in the early 1950s. New drugs and vaccines developed since the 1920s have also been strikingly effective against typhoid, whooping cough, poliomyelitis, measles, diphtheria, and tetanus; more recently great advances have been made in hormonal drugs, antihyper-tension drugs, antihistamines, anticoagulants, antipsychotic drugs, and antidepressants.[7]

Tables 1-7 and 1-8 illustrate the continuing influence of pharmaceutical products in lessened incidences of disease and death in the United States.[8] These statistics provide documen-tation for the impact of ethical drugs on public health, but only few data are available to quantify the additional importance of pharmaceuticals for private health. These private health benefits are often of considerable importance: the effects of anti-inflammatory agents on the functional capacity of arthritis patients, the implications of anti-anxiety and antidepressive drugs for patient quality of life, the cost savings of cimetidine in treatment of peptic ulcers. Nonetheless, the ordinary measures of public health produced by government agencies fail to capture these benefits.[9]

Offsetting these social benefits, there are clear social costs to pharmaceutical innovation. The complexity and diversity of patient reactions to ethical drugs restricts the abilities of con-sumers, their physicians, and often even pharmaceutical firms themselves to detect potential low incidence or long-term adverse side effects in the very potent drugs introduced since the thera-peutic revolution of the 1940s. It is by now well-established that laissez-faire policies under these market circumstances will result in distribution of pharmaceuticals whose risk is not fully appre-ciated, with occasional disastrous results. As a result of such social cost, national government regulation of product safety and distribution for pharmaceuticals has emerged in all the developed nations.

Unfortunately, safety regulation of the pharmaceutical industry presents its own social benefits and costs as well. In addition to reducing the frequency of adverse reactions and inappropriate therapies, contemporary regulations reduce the availability of and increase the delay and cost for new pharma-ceutical substances. Appropriate regulatory policy must strive to balance these social benefits and costs in order to insure the

TABLE 1-7   Reported Cases of Selected Diseases, 1951-1976

Diseases	1951	1960	1965	1976	Decline 1951-1976 (percent)
Measles (rubeola)	530,118	441,703	261,904	41,126	92
Meningococcal infections	4,164	2,259	3,040	1,605	61
Mumps	na[a]	na	152,109[c]	38,492	75 (from 1968)
Whooping Cough	68,687	14,809	6,799	1,010	99
Poliomyelitis	28,386	3,190	70	14	99
Rubella (German Measles)	na	na	45,975[d]	12,491	73 (from 1966)
Tuberculosis	85,607[b]	55,494	48,016	32,105	62
Typhoid Fever	2,128	816	454	419	80

[a]na = not available.
[b]1952 figure (1951 not available).
[c]1968 (not previously reportable).
[d]1966 figure.

SOURCE: U.S. Department of HEW, Public Health Service, Reported Morbidity and Mortality in the United States, 1976, Morbidity and Mortality Weekly Report, Vol. 25, No. 53 (Atlanta: Center for Disease Control, August 1977), p. 2; and U.S. Department of HEW, Public Health Service, Annual Reported Incidence of Notifiable Diseases in the United States, 1960, Morbidity and Mortality Weekly Report, Vol. 9, No. 53 (Atlanta: Communicable Disease Control, October 30, 1961), p. 4.

optimal use of medicinal products. In determination of this balance, polls repeatedly suggest that the American people are unwilling to make sacrifices in the safety and quality of ethical drugs simply to promote jobs and economic growth, and this panel explicitly endorsed this view. On the other hand, numerous reforms of U.S. FDA regulation have been proposed on purely medical grounds, to improve therapy for American patients, and the panel endorses many of these reforms. It is most important for the reader to recognize that any advancement of the economic position of U.S. pharmaceutical firms caused by these reforms is an explicitly and appropriately secondary reason for their adoption.

One important point, however, should be made: any balanced and appropriate policies toward the pharmaceutical industry should seek to sustain a large and rapid flow of truly safe and significant new drugs from American firms. It is precisely such balanced and appropriate policies that in the long run will most effectively advance both the public health and the competitive position of the U.S. pharmaceutical industry.

OVERVIEW AND LIMITATIONS OF THIS STUDY

The preceding has been an introduction to the U.S. and foreign pharmaceutical industries. Chapter 2 is an assessment of the

TABLE 1-8   Death Rate per 100,000 Population, 1920-1978

Cause of Death	1920	1940	1960	1978	Decline 1920-1978 (percent)
Tuberculosis, all forms	113.1	45.9	5.9	1.3	99
Dysentery	4.0	1.9	0.2	0.0[a]	100
Whooping Cough	12.5	2.2	0.1	0.0	100
Diphtheria	15.3	1.1	0.0[b]	–	100
Measles	8.8	0.5	0.2	0.0	100
Influenza and Pneumonia	207.3	70.3	36.6	26.7	87

[a]Bacillary dysentery and amebiasis.
[b]1959 (figures for 1960 and 1978 not available).

SOURCE: Ernst B. Chain, Academic and Industrial Contributions to Drug Research Nature (November 2, 1963) p. 441; and U.S. Department of HEW, Public Health Service, Health Resources Administration Final Mortality Statistics, 1978, Monthly Vital Statistics Report, Vol. 29, no. 6 (National Center for Health Statistics, Sept. 17, 1980).

competitive position of the U.S. pharmaceutical industry as it has developed since 1960. Chapter 3 evaluates the reasons for the current U.S. competitive position, Chapter 4 provides a brief discussion of new developments in the industry, and Chapter 5 offers options for public policy to strengthen the U.S. position internationally.

The focus of the report throughout is on the competitive position of U.S. pharmaceutical firms in the developed nations. This topic will be addressed directly, without extended discussion of the many peripheral issues that relate to health and medical care. Because many of these peripheral issues, however, are of considerable policy importance in their own right, it is useful to delineate some of them before consideration of the topic at hand.

Drug consumption in LDCs. The health concerns of less developed countries (LDCs) are different in many ways from those of Europe, North America, and Japan. Distinctive patterns of disease, widespread poverty, and illiteracy that reduce the efficacy of pharmaceutical treatments, and limited technical capacities of local regulatory officials all provide a unique set of concerns for the LDCs. These concerns have recently generated attempts by international institutions, notably the World Health Organization (WHO), to address the pharmaceutical-related medical problems of the LDCs through international regulation. While the LDC concerns and the WHO responses are of significance from the standpoints of world public health and international politics, the fact remains that LDC markets account for only a small minority of world ethical drug sales and virtually none of new drug introductions. Thus developments in these markets will have only limited impact on the relative competitive position of multinational drug firms. Further, an extreme paucity of data restricts any analysis of competition in the LDCs. For these reasons, this report examines exclusively the pharmaceutical markets and policies of the developed nations.

Genetic engineering. The biological production of chemical substances through genetically designed organisms offers exciting and eventually significant consequences for the pharmaceutical industry.[10] The short-term impact of this new technology, however, will be limited to particular market segments (vaccines and insulin) and major competitive effects will be delayed for as much as a decade or more. Such long-term technological developments were therefore beyond the scope of this study.

American health policy. Financial arrangements for the rapidly growing expenditures on health care for American citizens remains an area of controversy. Not only government policies toward Medicaid, Medicare, and hospital regulations, but also the policies of private insurers have been criticized for encouraging excessive consumption of health services. Suggested reforms in this area will indeed affect both U.S. and foreign firms, but consideration of U.S. health policy issues is beyond the scope of this study.

NOTES

1. Cited in Peter Temin, Taking Your Medicine, Drug Regulation in the United States, Harvard University Press, Cambridge, MA, 1980, p. 59.

2. Products of this era include thiamine, riboflavin, ascorbic acid, vitamin B_6 and vitamin B_{12}, along with thryoxine, testosterone, estrone, and progesterone. The discovery of cortisone also occurred in this period.

3. There are exceptions to the generally non-innovative character of OTC drugs. Fluoride toothpaste is one.

4. In recent years the growth of prescription drug sales has markedly slowed, largely due to the decreased frequency of new drug introductions. As a consequence, proprietary drug sales may now grow more rapidly than prescription drug sales.

5. For additional discussion of the competitive structure of the U.S. pharmaceutical industry, see Office of Technology Assessment, Patent Term Extension and the Pharmaceutical Industry, U.S. Government Printing Office, Washington, D.C., 1981, pp. 16-19.

6. Charles River Associates, "The Effects of Patent Term Restoration on the Pharmaceutical Industry," Boston, MA, May 4, 1981, (report of OTA) pp. 17 and 74.

7. Victor Fuchs, Who Shall Live?, Basic Books, New York, 1974.

8. New drugs are not the sole cause of recorded declines in mortality and morbidity. Improvements in sanitation, education, and income have also contributed substantially.

9. For a brief discussion of the nature and significance of private health benefits, see William Hubbard, "Defining the Role of Medicinals in Health," presentation to the Tenth IFPMA Assembly, October 1980.

10. For a discussion of the long-run effects of genetic engineering on the pharmaceutical industry, see Office of Technology Assessment, Impacts of Applied Genetics, U.S. Government Printing Office, Washington, D.C., 1981.

2
Competitive Position of the U.S. Pharmaceutical Industry

A fundamental charge for this study is to assess the competitive position of U.S. pharmaceutical firms against their major foreign counterparts. Three complexities immediately beset the panel's efforts to execute this assessment:

1) The extensive and increasing multinational diffusion of individual pharmaceutical firms has rendered "U.S. pharmaceutical industry" a term of unclear meaning. The larger pharmaceutical houses founded in America have long since developed extensive facilities in dozens of foreign markets. Conversely, foreign-based firms have established operations in the United States; in fact, the largest U.S. firm in the mid-1970s in terms of pharmaceutical sales to American consumers was Roche Laboratories, a subsidiary of the Swiss-based firm Hoffman LaRoche. The widespread practices of licensing innovations, marketing agreements, and joint ventures among firms of many nationalities further complicates the assignment of specific facilities and specific products to individual nations.

2) The "competitive position" of firms in an industry that exhibits rapid growth of markets and radical product innovation is a multidimensional phenomenon that is not easily characterized. From one perspective, current rates of return are an overall summary measure of competitive position. Yet, these returns actually appraise past corporate performance and achievements rather than indicate future industrial strength. From a second perspective, current market shares provide a reasonable proxy for competitive position in the immediate future. For the longer horizon, however, the intensely innovative nature of the pharmaceutical industry makes the extent and vitality of corporate research a crucial determinant of competitive success. Reduction of these and other dimensions of competitive position into a single univariate index is in no way a simple task.

21

3) Finally, the charge to "assess" the pharmaceutical industry presumes a coherent perspective for evaluation. Yet, several substantially varying perspectives immediately present themselves. American labor will assess the pharmaceutical industry on the basis of the number of jobs and the volume of salaries generated domestically, investors on the basis of future profits, and consumers on the basis of variety, safety, effectiveness, and costliness of remedies. From a broader national perspective, the level of export earnings, the industrial concentration of output, and the level of long-run expenditures for national health care are factors which must validly be considered in assessment. Difficult choices are faced in reducing these potentially conflicting goals into a single "public interest."

The strategy of this report for coping with these complexities is as follows. Six aspects of industrial performance are considered: research effort, innovational output, production, sales, market structure, and international trade. Relevant data for these six aspects are reported for the post-1960 era for pharmaceutical institutions aggregated in two ways: first, by country of location, and second, by country of ownership. Thus, for purposes of this report, "U.S.-located firms" refer to all pharmaceutical facilities that physically operate within the territorial boundaries of the U.S. regardless of national ownership, while "U.S.-owned firms" refer to the pharmaceutical facilities of the U.S.-based multinational firms regardless of their geographic locations. In many cases, data limitations allow only one of these two aggregations. In other cases, common sense dictates that only one definition of nationality be used; export data necessarily refer to pharmaceutical activities within national boundaries, while market share data necessarily refer to sales of multinational firms owned by the same country (e.g., the U.S. share of the Japanese market). Each aggregation is important, though for different purposes. Aggregations by country of location enable comparison of the different economic experiences and public policies of various national governments and how these affect the pharmaceutical industry. Aggregations by country of ownership enable evaluation of differing national management strategies and modes of industrial operation.

However the issue of nationality is settled, the relative position of U.S. firms has been at best stable and has at worst deteriorated with regard to each of the six criteria considered. In other words, the U.S. share of world pharmaceutical research, innovation, production, sales, and exports and the number of U.S. firms that are active participants in the ethical drug markets have all been constant or declined since 1960; in some instances, this decline has been dramatic. The unidirectional nature of these trends somewhat relieves the second and third difficulties raised above. Since all of the chosen measures indicate stability or decline of the American competitive position, complex problems of the relative importance of each measure are minimized.

It is important to realize that any decline of American firms discussed in this report is relative to their foreign counterparts and not absolute. For example, during the 1970s, levels of production and research for pharmaceutical facilities within the territorial U.S. gradually increased. Yet, during this same period, production and research expenditures increased extremely rapidly abroad. As a consequence of these differing growth rates, the U.S. share internationally of both research expenditures and production markedly declined.

RESEARCH

Research is the foundation of competitive strength for modern pharmaceutical firms. As shown earlier, growth in sales and profits for major ethical drug companies are derived from a handful of commercially successful new products discovered and developed through industry research efforts.

Pharmaceutical research may be divided into four phases:

1) Basic research--advancement of basic pharmacological knowledge. This is the only phase not directly regulated by government, although government regulation has a substantial indirect impact. About 12 percent of the pharmaceutical research performed in the United States is basic.[1]

2) Discovery effort--the synthesis of active substances and the establishment of biological effect.

3) Applied research--the extensive biological (animal) and clinical (human) testing of substances to determine pharmacological activity and risk of adverse effects.

4) Development--the determination of dosage form, the development of manufacturing processes, and the production of drug product.

Pharmaceutical research is characterized by substantial risks and lengthy time requirements. For research that will lead to completely new products, the process begins with assemblage of a research team to consider a therapeutic problem, to review the literature, to examine hundreds of chemical substances, and to select a handful of these substances for further investigation. The chosen substances or potential drug candidates will be tested in animals for pathological and toxic effects. Only about 2 percent of those compounds tested biologically will be subsequently tested in humans, although the attrition rate varies enormously across different therapeutic fields.[2] Most compounds will fail to demonstrate suitable therapeutic advantages, or will not be commercially promising. Two to four years on average will elapse from the selection of a potential drug candidate to the initiation of human testing.

Once the stage of clinical testing is reached, regulatory review of the research design is required in many nations--in the United States, an Investigational New Drug (IND) exemption is required; in the United Kingdom, this requirement was labeled the Clinical Trial Certificate (CTC). In early 1981 the U.K. CTC was replaced by the Clinical Trial Exemption procedure, which is now quite different from the U.S. IND.

Clinical testing under the IND proceeds in three phases. In Phases I and II, healthy volunteers are administered the drug to examine basic pharmacological effect and safety, and a limited number of patients receive the drug to examine its efficacy in treatment of a specific illness. Expanded studies are conducted in Phase III to confirm the findings of Phase II and to uncover uncommon adverse reactions.

After the first three phases of clinical trial are completed, the compound is submitted to the regulatory authority for permission to market the drug. In the United States, this submission is entitled the New Drug Application (NDA). Only about 10 percent of those drugs that are initially included in clinical trials will subsequently be the subject of an NDA. Average total time for the IND/NDA period of testing and approval in the United States is currently in excess of eight years. As of 1976 the mean duration of the IND/NDA period for New Chemical Entities (NCEs) self-originated by U.S.-located firms was in excess of nine years. However the mean duration of the IND/NDA period for acquired NCEs was about 4.5 years.[3] In short, the full period from initiation of basic research into a particular pharmacological problem to the commercial launching of a new product may exceed 15 years. Recently, an additional Phase IV of studies have been required on consumers of a few drugs after marketing.

Research expenditures by pharmaceutical firms have substantially increased during the past two decades, but at greatly divergent rates among facilities. Table 2-1 presents basic data on expenditures for pharmaceutical R&D by corporate facilities aggregated by national location. While there are inevitable complications for interpretation caused by exchange rate fluctuations, it is clear that growth rates for such R&D have been significantly higher for facilities in Western Europe and Japan than in the United States. More recent data are presented in Table 2-2 and indicate that higher rates of growth have persisted

TABLE 2-1 Pharmaceutical R&D Expenditures by Nationality (Location) 1964, 1973, and 1978

	1964		1973		1978	
	Level ($ million)	World Share (percentage)	Level ($ million)	World Share (percentage)	Level ($ million)	World Share (percentage)
United States	282	60	640	34	1,159	28
Germany	40	8	310	16	750	18
Switzerland	38	8	244	13	700*	17
Japan	27	6	236	13	641	15
France	28	6	166	9	328	8
United Kingdom	29	6	105	6	332	8
Italy	15	3	82	4	147	4
Sweden	9	2	33	2	72	1
Netherlands	9	2	26	1	72*	1

*Estimated.

NOTE: Data are in current dollars and represent expenditures for both human and veterinary research.

SOURCE: OECD, Directorate for Science, Technology, and Industry, *Impact of Multinational Enterprises on National Scientific and Technical Capacities*, mimeograph, Paris, 1977. 1978 data from Table 2-2.

TABLE 2-2 Pharmaceutical R&D Expenditures by Nationality (Location),
Recent Years

Year	United States	United Kingdom	Federal Republic Germany	Japan	France	Italy
1973	906	67	950	87.5	941	73.9
1974	857	64	942	81.8	789	63.5
1975	893	87	1,000	95.3	1,021	73.7
1976	927	102	1,148	–	1,073	na
1977	947	106	1,243	112.9	1,138	na
1978	968	124	1,389	129.3	–	79.6
1979	967	na	na	–	–	
Annual rate of growth	1.1%	13.1%	7.9%	8.1%	4.8%	1.5%

NOTES: Data are in millions (except for Japan and Italy, in billions) of constant (1975 base) local currency and represent expenditures for both human and veterinary research.

Deflator is the wholesale price index in each country as compiled by the International Monetary Fund.

SOURCES: Pharmaceutical Manufacturers Association, *Annual Survey Report*, PMA, Washington, D.C., various years.

Association of the British Pharmaceutical Industry, *Annual Report*, ABPI, London, various years.

Bundesverband der Pharmazeutischen Industrie, *Pharma Jahresbericht*, BPI, Frankfurt, various years.

Droit et Pharmacie, "Research," June 1980.

International Monetary Fund, *International Financial Statistics Yearbook*, IMF, Washington, D.C., 1979.

at least for Japan, West Germany, and the United Kingdom. It is clear from these data that the share of world pharmaceutical research that is located in the United States has fallen from about two-thirds in the early 1960s to about one-third today.

The U.S.-owned share of world pharmaceutical R&D expenditures may be marginally larger than the U.S.-located share, as U.S. multinational pharmaceutical firms appear to spend more for research abroad than do foreign-owned firms in the United States. Reports from U.S.-owned multinationals indicate that the foreign subsidiaries of these firms spent $117 million for research in 1973 and $238 million in 1978, or approximately a constant 6 percent share of world expenditures.[4] Thus, note by way of example that if foreign-owned firms conducted absolutely no pharmaceutical R&D in U.S.-located laboratories, then the U.S.-owned share of ethical drug R&D would be simply the U.S.-located share (given above) plus 6 percent. The U.S. figures plus 6 percent thus provide an upper bound on the U.S.-owned share of world R&D. However foreign-owned firms do maintain large research facilities in the United States, though, unfortunately, the exact division of U.S.-located R&D between that of U.S.-owned and foreign-owned firms is not available. In any case, industry consensus indicates that, although the foreign-owned share of U.S.-located pharmaceutical R&D has not dramatically changed, if anything it has slightly increased. Hence, while the trend in the U.S.-owned share of world pharmaceutical R&D cannot be exactly estimated, it is clear that this share has markedly dropped.

In sum, while U.S.-owned expenditures for pharmaceutical research at home and abroad are large and growing, they have not increased nearly enough to match the exceptional expansion of foreign-owned research efforts. The upshot, measured by either location or ownership, is a significant decline in the U.S. share of R&D, the foundation of competitive position in this industry.

INNOVATION

The enormous increase in world pharmaceutical R&D expenditures might be expected to yield a comparable surge of new products for consumers. Unfortunately, levels of innovative productivity in the industry, at least as measured by the number of NCEs brought to the market, have been, at best, stable for the last two decades and have sharply dropped since the 1950s. Figure 2-1 demonstrates these trends for the United States. Although the medical or therapeutic value of today's NCEs is probably better than in the past, it is a straightforward conclusion that the average cost per innovation has drastically risen in the last 20 years. An overview of six economic studies that examined the increased costs of pharmaceutical innovation found the cost per NCE to have risen in constant (1980) dollars from approximately $6.5 million before 1962 to about $44.7 million in 1980 (excluding the cost of capital). The average R&D expenditure per NCE (including capital cost) has been estimated by the most prominent of these six studies at $70 million in 1980 dollars.[5]

The fundamental reason for the dramatic increase in innovation costs lies in the substantially greater clinical trials and toxicology testing performed in the process of bringing a new compound to market. Advances in medical science have vastly improved the abilities of pharmaceutical researchers to identify potential adverse reactions and to predict therapeutic efficacy. While most of these costly procedures have been mandated by national regulatory authorities, some would have been adopted by industry in any case.

The decline in NCE introductions is thus not totally indicative of a decline in basic pharmaceutical innovation. Indeed, patent filings in the United States would indicate that basic innovation has increased in pace with increased research expenditures. Patent filings by U.S. firms have roughly doubled since 1963, while filings of foreign firms have quadrupled. Instead, the costly expense of premarket testing has forced firms to be much more selective of those compounds to be brought to market. Fewer compounds will possess sufficient market potential to recoup the substantial and increasing R&D costs incurred for each marketed substance. One indication of this greater selectivity is the decline in the ratio of INDs filed to patents granted, which is now at half of its level in 1963 (see Table 2-3). While the number of compounds entering clinical testing is not observable prior to the 1962 imposition of the IND requirement, there is every reason to believe that the 1950s equivalent of this ratio was even higher.[6]

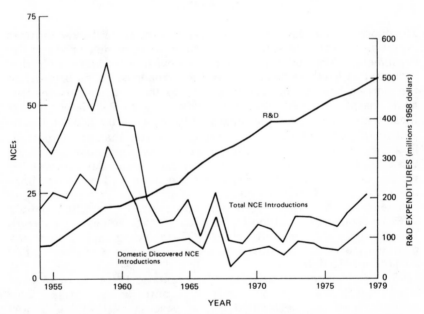

FIGURE 2-1 Domestic U.S. Introductions and Discoveries of New Chemical Entities (NCEs) and Pharmaceutical R&D Expenditures, U.S.-based Firms, 1955-1979

NOTE: R&D figures exclude veterinary efforts but include overseas expenditures of U.S.-based firms.

SOURCE: Henry Grabowski, " Public Policy and Innovation: The Case of Pharmaceuticals," Technovation, 1982.

TABLE 2-3   Total U.S. Patent Registrations, Drugs and
Medicines, and Total U.S. IND Filings, 1963-1977

Year	Patents	INDs	IND-Patent Ratio
1963	1,532	1,066	0.69
1965	1,865	751	0.40
1967	2,438	671	0.28
1969	2,630	956	0.36
1971	2,417	923	0.38
1973	3,166	822	0.26
1975	4,385	876	0.20
1977	4,168	925	0.22

SOURCES: (Patents) U.S. Patent and Trademark Office, Office
of Technology Assessment and Forecast, *Active Patent Classification in R&D Intensive Industries and Fifty-two Standard Industrial Classification Categories*, U.S. Government Printing
Office, Washington, D.C., 1976.

(INDs) Pharmaceutical Manufacturers Association, *Prescription Drug Industry Factbook 1980*, PMA, Washington, D.C.,
1981.

The upward trend in costs of innovation is of course an
international phenomenon that has led in all industrial nations to
comparable extensive pretesting and selectivity in pursuit of new
drugs. The inevitable consequence has been a worldwide decline in
introduction rates (see Figure 2-2). While research costs have
risen in all countries, the increase has apparently been higher in
the United States than elsewhere.[7] When the greater expense of
innovation in the United States is considered alongside of relatively decreasing U.S. levels of research expenditure, it is not
surprising to find that the U.S. share of pharmaceutical innovation
has dropped over the last two decades--in other words, that
foreign levels of innovation have declined less severely since the
1950s than those of the United States.

Relative national success with pharmaceutical innovation may
be documented at three distinct points during the innovation
process: patent filing, IND filing, and actual introduction--due to
the fact that data are systematically collected at these points.
Each of these three sets of statistics presents advantages and
disadvantages for use as indicators of contemporary competitive
advantage. Because of the lengthy time lag between discovery
and marketing, data on currently introduced drugs will be indicative of economic conditions and management decisions of as much
as a decade ago. The introduction data are, however, available for
most major national markets. IND filings and patents issued will
more nearly reflect current circumstances, but are readily
available only for the United States. While there is little reason

to expect trends in these filings for the United States (the world's largest market for pharmaceuticals) to be unrepresentative of worldwide conditions, it would have nonetheless been useful to have corroborating evidence from other nations.

Turning first to the U.S.-owned share of drugs actually marketed, data on NCEs introduced over the past few decades are given in Table 2-4 for the United States and in Table 2-5 for the world. Both tables demonstrate stability in the U.S. share of introductions, except for a downturn around 1970 in the U.S.-owned share of introductions into the United States. This temporary downturn (or increase in foreign-owned share) in the United States is also illustrated in Figure 2-3.

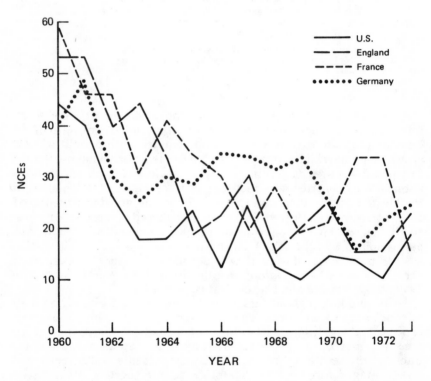

FIGURE 2-2 Annual Marketing of NCEs in the United States, England, France, and West Germany

SOURCE: Compiled from data of Paul de Haen and presented as part of FDA Commissioner Alexander Schmidt's testimony before Senate Subcommittee on Health of Committee on Labor and Public Welfare, 1974.

TABLE 2-4   NCEs Marketed in U.S. by Year of First Introduction and Nationality (Ownership) of Innovating Company, 1951-1980

Time Interval	Number of NCEs	United States		Foreign	
		Origin	Percentage	Origin	Percentage
1951-56	172	109	63	63	37
1957-62	188	109	58	79	42
1963-68	88	53	60	35	40
1969-74	76	37	49	39	51
1975-80	94	54	57	40	43

SOURCE: Center for Study of Drug Development, University of Rochester.

Quite different findings emerge from examination of data that are collected for an earlier stage in the innovation process and thus are more representative of the contemporary economic environment for pharmaceutical research. Patent data are given in Table 2-6 and show a drop in the U.S.-owned share of drug patents from 65 percent to about 50 percent during the 1970s. Breakdowns for IND filings in the United States are given in Table 2-7 and in Figure 2-4 and require a brief word of explanation. INDs in Table 2-7 are given by country of ownership, defined here on the basis of the firm holding patent rights, and not necessarily the firm actually marketing the NCE. An example of this difference is provided by Motrin, an extremely successful drug discovered and developed by the British firm Boots, but marketed under license in the United States by Upjohn. INDs in Figure 2-4 are given in terms of nationality by location, indicating the country in which synthesis physically occurred regardless of ownership of facilities housing this research. All IND data count only original filings for new chemical entities. Both Table 2-7 and Figure 2-4 show a continued decline in U.S.-owned or located INDs alongside rough stability in levels of foreign INDs. The comparative trends in levels (downwards vs. stable) that are visible in Figure 2-4 are even more revealing in this case than the simple percentages.

In conclusion, the sharp decline in the U.S. share of world pharmaceutical R&D expenditures in the 1960 to 1970 period was followed by a significant drop after about 1967 in the U.S.-owned share of medicinal patents filed in this country and a continued decline after 1960 in the U.S. share of NCEs started in American clinical trials. By the end of the 1970s, no comparable decline in the U.S.-owned share of marketed NCEs had occurred and indeed that share remained at levels prevailing since the 1960s. Continued stability in this share is at best uncertain.

TABLE 2-5   NCEs Marketed Worldwide by Year of First Introduction and
Nationality (Ownership) of Innovating Company, 1961-1977 (percentages)

	1961-1964	1965-1969	1970-1974	1975-1977
United States	24.5	22	23	24.5
France	17.5	22	19	12.5
West Germany	16	11.5	8.5	14.5
Japan	9	10	10	9
Switzerland	9	6	7	6
Italy	5	7	6.5	11
United Kingdom	6.5	5	3.5	7
Others	28.5	16.5	22.5	15.5
Total NCEs	353	410	377	190

SOURCE: Erika Reis-Arndt, "New Pharmaceutical Entities, 1961-1977," *Die Pharmazeutishen Industrie*, Vol. 40, Nr. 11, 1978. (Translation by A. M. Lee and C. M. Sonne.)

PRODUCTION

The levels of pharmaceutical production located in the United States have been roughly stable for the mid-1970s, while production located in Western Europe and Japan has exhibited substantial growth. Between 1965 and 1975, U.S.-located production grew at a 5 percent annual rate compared to a 15 percent rate abroad. The inevitable result of such divergent patterns of growth is a falling U.S.-located share of world pharmaceutical production. In the early 1960s, U.S. production was nearly twice the value of Western Europe output; today that statistic is effectively reversed with U.S.-located output at less than 60 percent of the European total. Over the same period, production located in Japan grew from one-third that of the United States to 75 percent of the U.S. level.[8] For comparative production data see Table 2.8.

The relative decline of production by U.S.-owned firms is expectedly much less severe than that of U.S.-located establishments due to the substantial increase in production abroad by the former. For U.S.-owned firms, overseas production increased from $0.7 billion in 1963 to over $6 billion in 1978. Growth rates of overseas production for U.S.-owned firms thus substantially exceeded those of domestic production, insuring that by 1978 foreign production accounted for 40 percent of the U.S multinational total. Comparable figures for other nations are not available.

SALES

The commercial significance of innovation for pharmaceutical firms lies in the extreme importance of new products for overall

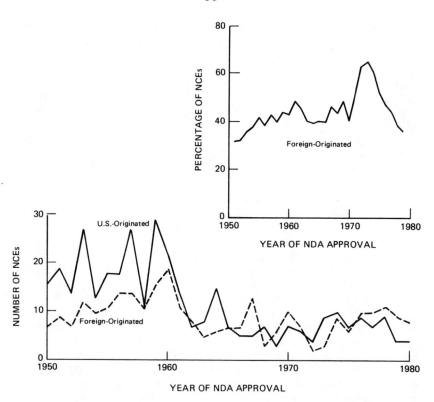

FIGURE 2-3 Stratification of U.S.-Marketed NCEs by National Origin

NOTE: Number of U.S.-approved NDAs marketed in the U.S. between 1950 and 1980 by year of NDA approval. The data were stratified by the source or national origin of the NCEs based on questionnaire responses made by approximately 92 percent of the pharmaceutical firms located in the United States. National origin refers to the nationality of the firm that originally synthesized, owned, and/or developed the NCEs. U.S.-originated NCEs were therefore NCEs that were either self-originated by U.S.-owned firms or by foreign firms. Similarly, foreign-originated NCEs were NCEs that were either self-originated by foreign-owned firms or licensed/acquired from foreign forms by U.S.-owned firms or by other foreign firms. In the upper insert, the percentage contribution of foreign-originated NDA approvals is described as a three-year moving average by year of NDA approval.

SOURCE: Center for the Study of Drug Development, University of Rochester.

levels of sales, and hence earnings. Firms that fail to introduce new drugs will find their sales growing slowly, if at all, and ultimately declining. Another way of expressing this fact is that patented drugs experience a "product life cycle." After introduction, the medical community gradually adopts the new drug and sales rise while market share for the drug increases. Eventually, however, newer and superior drugs with similar therapeutic functions will be introduced, and sales for the original drug will slow while its market share declines. Thus a pharmaceutical firm without new products will in due course hold a portfolio of marketed drugs all on the downward side of their product life cycle.

The extent of volatility in pharmaceutical sales due to innovation may be seen in Table 2-9, which lists the best selling drugs in terms of domestic U.S. sales for several recent years. Examples of the product life cycle are provided in this table by Keflin (chronologically ranked since 1970--6, 5, and 22), Mellaril (chronologically ranked 18, 8, 14, and 25), and numerous other drugs. Perhaps the clearest indication of the sales impact of new pharmaceutical products is provided by the fact that only 4 of the 30 top-selling products in 1965 remained in the top 30 by 1980. In light of this importance of innovation for pharmaceutical sales, it is relatively unsurprising that the U.S.-owned share of pharmaceutical sales has to a large extent followed trends in the U.S.-owned and U.S.-marketed shares of NCE introductions. For example, the surge between 1968 and 1978 of the foreign share of drugs marketed in the United States (presented in Figure 2-4) is associated with a drop in the U.S.-owned firms' share of U.S.-located pharmaceutical sales of all drugs (as shown in Table 2-10). By 1979, the market share of U.S.-owned firms had not yet turned upwards despite the pronounced recovery in the U.S. share of NCE introductions beginning in the mid-1970s. If the U.S. share of

TABLE 2-6   U.S. Patent Registrations, Drug and Medicines, by Year of Filing and Nationality (Ownership) 1963-1979 (percentages)

Year	United States	West Germany	Japan	Switzerland	United Kingdom	France	Italy	Other
1963	66	8	2	6	3	3	3	9
1965	64	7	4	5	4	4	3	9
1967	65	6	5	5	5	5	2	7
1969	62	7	5	6	5	4	2	9
1971	60	9	6	6	4	6	2	7
1973	58	11	6	6	5	5	2	7
1975	57	11	8	4	5	6	2	7
1977	51	13	8	4	8	6	3	7
1979	50	12	9	5	8	6	2	8

SOURCE: U.S. Patent and Trademark Office, Office of Technology Assessment and Forecast.

TABLE 2-7   U.S. INDs Filing, NCEs Only, by Nationality (Ownership) of Innovating Company, 1965-1979

Time Interval	Number of NCE - INDs	United States		Foreign	
		Origin	Percentage	Origin	Percentage
1965-69	397	287	72	110	28
1970-74	339	238	70	101	30
1975-79	223	126	56	97	44

SOURCE: Center for the Study of Drug Development, University of Rochester.

NCE introductions remains at current levels, then short-run upturn in market share might be viewed as possible for U.S. firms.

In a similar vein, the effective stability of the U.S.-owned share of worldwide NCE introductions (presented earlier in Table 2-5) is associated with rough stability in the U.S.-owned share of pharmaceutical sales in France, West Germany, Italy, and Japan. Only one anomaly arises here in that the U.S.-owned share of British-located drug sales dropped significantly in the mid-1960s. An explanation at least in part for this occurrence derives from the policy change at that time by the British Public Health Ministry, which directly pays for most pharmaceutical sales in that country, to by fiat lower the price (hence sales volume in pound terms) for several best-selling antibiotic products. Many of these products were marketed by American firms.

Even more significant and interesting than the U.S.-owned share of all drug sales is the comparable share of top-selling drugs. The market share data for all drugs in Table 2-10 cover numerous generic and off-patent drugs for which profit margins are extremely low. As discussed in the introduction, however, the bulk of sales and earnings for each pharmaceutical firm are derived from a handful of very successful drugs. Thus the U.S.-owned share of these best selling drugs is an important and superior measure of the U.S. share of innovational earnings. And it is precisely these earnings that pay for R&D costs to develop future NCE introductions. Returning to Table 2-9, note that foreign-owned drugs (which are in some cases marketed by U.S.-owned firms) are denoted with an asterisk. Aggregating the sales of foreign-owned drugs in Table 2-9 gives the foreign-owned share of sales in the top 15 and top 30 selling drugs for various years:

Percentage Foreign-owned

	1965	1970	1975	1980
Top 15	36.8	59.6	48.7	34.4
Top 30	31.0	46.4	43.1	35.4

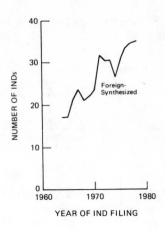

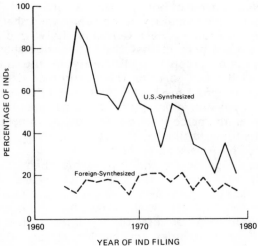

FIGURE 2-4 Stratification of U.S.-Filed INDs by Nationality of Parent Company

NOTE: Number of INDs filed in the United States between 1963 and 1979. The data were stratified by nationality of the parent company synthesizing the NCEs based on questionnaire responses made by about 95 percent of the U.S.-owned firms and about 70 percent of the foreign-owned subsidiaries located in the United States. In the upper insert, the percentage contribution of NCEs synthesized by foreign firms is described as a three-year moving average by year of IND filing.

SOURCE: Center for the Study of Drug Development, University of Rochester.

TABLE 2-8   Annual World Production of Pharmaceutical Products, 1968-1978 (percentages)

	1968	1970	1972	1974	1976	1978
United States	38.0	35.0	33.0	29.0	30.0	27.0
Japan	13.0	14.5	14.5	16.0	16.0	20.0
West Germany	8.5	9.0	10.0	10.5	10.0	10.0
France	6.0	5.5	6.0	6.5	6.5	6.5
United Kingdom	6.0	5.5	5.5	5.5	5.0	5.5
Italy	5.0	4.5	4.5	5.5	5.0	4.5
Switzerland	2.0	2.0	2.0	2.5	2.5	3.0
Others	21.5	24.0	24.5	24.5	25.0	23.5

SOURCE: Bundesverband der Pharmazeutischen Industrie, *Pharma Jahresbericht*, BPI, Frankfurt, various years.

The U.S. share of top-selling drugs thus very closely followed trends in U.S.-marketed NCE introductions (as per Figure 2-4). Given the importance of the U.S. market to U.S.-owned firms, these data indicate that these firms suffered a brief but severe deterioration in competitive position in the early 1970s in terms of relative sales and earnings.

Comparison of worldwide sales for major pharmaceutical firms is made possible by a profile of the international industry provided for this study and listed in Table 2-11. Only the largest firms are included in this profile, and they are grouped by nationality of ownership. While fluctuations in exchange rates over time make comparison of national growth rates a difficult exercise, it is clear that the sales of American-owned firms have grown in recent years at roughly the same rate (13.1 percent annually) as those of most foreign firms. This similarity in growth rates is of course reflected in the stability in the market shares of nationally located sales for U.S.-owned firms and is based on the extended stability in the U.S.-owned share of world introductions of NCEs.

Two examples of the importance of innovation for sales growth can be seen from the industry profile. From a positive perspective, the spectacular growth of sales for the American firm, Smith Kline, is due largely to a single drug, Tagamet, introduced in 1977. On the other hand, Warner-Lambert has not had introductions of significant success in recent years, and its growth rates for pharmaceutical sales have badly lagged those of the market as a whole.

STRUCTURE

Two basic changes in the structure of the world pharmaceutical industry have evolved during the past two decades--greater

TABLE 2-9 Ethical Drugs—Purchases of Top 30 U.S.-Located Trademarks (millions of dollars)

	1980		1975		1970		1965	
1.	Tagamet	233	*Valium[1]	273	*Valium[1]	125	*Librium[1]	59
2.	*Valium[1]	221	Darvon Prods.	91	*Librium[1]	86	Chloromycetin	42
3.	*Inderal[3]	179	Aldomet	81	Darvon Prods.	77	Darvon Prods.	41
4.	*Motrin[3]	135	*Lasix[4]	70	*Orinase[13]	43	*Orinase[13]	40
5.	Aldomet	133	Keflin	70	*Polycillin[14]	42	*Thorazine[15]	31
6.	Dyazide	133	Keflex	69	Keflin	41	Declomycin	30
7.	Keflex	131	*Librium Prods.[1]	62	Indocine	39	*Valium[1]	27
8.	Clinorile	115	Premarin Prods.	57	*Mellaril[7]	37	Equanil	25
9.	*Lasix[4]	108	Indocin	55	Premarin	37	Terramycin	21
10.	Ovral Prods.	97	*Motrin[3]	54	Erythrocin	30	Pan Alba	20
11.	Ortho-Novum	84	Ortho-Novum	54	Maalox	29	Achromycin	19
12.	Darvon Prods.	84	Garamycin	51	Ortho-Novum	29	Enovid	19
13.	Indocine	75	Ovral Prods.	48	*Thorazine	29	Stelazine	18
14.	Maprosyn	75	*Mellaril[7]	44	Ilosone	28	Ortho-Novum	17
15.	Tylenol w/cod.	64	*Elavil[10]	43	*Lasix[4]	28	Diuril	17
16.	Mandol	63	Triavil	40	Triavil	27	Peritrate	16
17.	Garamycin	63	Aldactazide	39	*Loridine[16]	27	*Erythrocin	16
18.	Diabinese	63	Dyazide	37	Achromycin	25	*Mellaril[7]	15
19.	Cleocin	59	Aldoril	36	Stelazine	23	Mysteclin-F	15
20.	*Persantine[5]	59	Maalox Prods.	35	Peritrate	23	Premarin	15
21.	Aldoril	59	Medrol Prods.	34	Pavabid	22	Dexamyl	15
22.	Keflin	54	Diabinese	32	V-Cillin K	21	Maalox	14
23.	Isordil	54	*Inderal[2]	32	Equanil	20	Ilosone	14
24.	*Slow-k[6]	52	*Empirin[17]	32	Diabinese	20	Enovid-E	14
25.	*Mellaril[7]	51	Mylanta	30	*Elavil[10]	20	*Fluothane[2]	14

No.								
26.	*Hygroton[8]	50	Pavabid	29	*Fluothane[2]	19	Lincocin	14
27.	Erythrocin Prods.	49	*Tofranil/PM[11]	29	Ornade	19	Pentids	14
28.	*Zyloprim[9]	49	*HydroDIURIL[12]	28	Talwin	19	Furadantin	13
29.	Vibramycin	47	Ilosone	28	*HydroDIURIL[12]	18	V-Cillin K	13
30.	*Dalmane[1]	47	*Ser-Ap-Es[6]	28	Decadron Prods.	18	*HydroDIURIL[12]	13

*

Patent Current Status	Patent(s) Owned by	Marketed in U.S. by
[1] (Librium expired)	Roche (Switzerland)	Roche
[2]	Imp. Chemical Ind. (U.K.)	Ayerst (U.S.)
[3]	Boots (U.K.)	Upjohn (U.S.)
[4] (expired)	Hoechst (Germany)	Hoechst
[5] (expired)	Boehringer Ingelheim (Germany)	B-1
[6]	Ciba-Geigy (Switzerland)	Ciba
[7]	Sandoz (Switzerland)	Sandoz
[8] (expired)	Ciba-Geigy (Switzerland)	USV (U.S.)
[9] (licensed)	Wellcome Foundation (U.K.)	Burroughs-Wellcome
[10] (licensed)	Roche & Merck	MSD (U.S.)
[11] (expired)	Ciba-Geigy (Switzerland)	Geigy
[12] (licensed)	Ciba-Geigy (Switzerland)	MSA (U.S.)
[13] (expired)	Hoechst (Germany)	Upjohn (U.S.)
[14] (licensed)	Beecham (U.K.)	Bristol (U.S.)
[15] (expired)	Rhone-Poulenc (France)	SKF (U.S.)
[16]	Glaxo (U.K.)	Lilly (U.S.)
[17]	no patent	Burroughs-Wellcome

SOURCE: Merck & Co., Inc., MSD Strategic Planning and MSD Marketing and Sales Research, West Point, PA. Original data from IMS, Inc., Ambler, PA.

TABLE 2-10   Market Share of U.S.-Owned Multinational Pharmaceutical Firms, Selected National Markets, Selected Years

Market	U.S.-Owned Firms' Market Share			
	1965	1970	1975	1979
World	na	na	31.9	29.3
United States	86.9	83.5	80.8	79.8
United Kingdom	45.9	39.5	37.7	36.6
West Germany	na	na	17.4	18.0
France	na	na	18.3	18.6
Italy	na	na	18.0	18.3
Japan	na	na	7.0	6.7

SOURCES: (1975 and 1979—except U.S.): Eli Lilly and Company, Corporate Economic Staff, 1980. Original data from IMS, Inc., Ambler, PA.

(1965 and 1970, U.K.): National Economic Development Office, *Focus on Pharmaceuticals*, HMSO, London, 1972.

(U.S. data): Merck & Co., Inc. MSD Strategic Plan, 1981

Original data from IMS, Inc., Amber, PA.

concentration of innovation among larger firms and increased internationalization of the industry. Both trends have important implications for the competitive status of U.S. pharmaceutical firms.

The effects of sharply rising costs of innovation have not been evenly distributed among all drug firms. There is growing evidence that the substantially more extensive and costly testing and the years of delay between synthesis and marketing of a drug have combined to make costs and risks of innovation particularly onerous for firms with small research budgets. Historically, both in the pharmaceutical industry and in other sectors of the economy, the costs per innovation within a given sector have not systematically differed on the basis of firm size. In economic terms, no economies of scale existed. When the costs per innovation were roughly comparable among different sizes of research units, those firms conducting research on a smaller scale (beyond some obvious minor threshold) suffered no disadvantage in their abilities to innovate against and compete with larger firms. Unfortunately, recent shifts in costs and risks appear to have especially affected smaller pharmaceutical firms, with the result that costs per introduced NCE appear to decline with size of total research effort. In other words, economies of scale persist until there is some large scale of research operations entailing simultaneous study of numerous promising NCEs.[9] Current industry estimates of the annual level of research expenditures at which average innovational costs cease to decline range as high as $100 million a year. Only very few firms maintain annual research budgets of that magnitude. The higher costs per innovation for small firms render small-scale research operations relatively less productive per dollar spent and hence less profitable. An unsurprising consequence of this development has been the declining significance of smaller firms in the pharmaceutical innovation process. This trend has particularly serious consequences for nationally owned pharmaceutical industries based on smaller and medium-size firms, such as those in Belgium and Sweden.

Data on the concentration of sales and innovation for the pharmaceutical industry shed light on this issue. The aggregate market share of the 4 largest, 8 largest, and 20 largest U.S.-located and British-located firms are provided in Table 2-12. From the vantage of these sales data, the industry appears to be relatively unconcentrated with 20 or more active participants in the market still accounting for only 75 percent of sales. Further, only very subtle changes are visible from these data as to any shifts in concentration of the industry. Data on concentration of innovation given in Table 2-13, however, tell a different story, at least for the United States. The largest U.S.-located firms amount for a large and growing share of U.S.-located new drug sales. Greater concentration of new drug sales indicates that smaller firms are failing to innovate as rapidly or successfully as larger firms, a finding also indicated by the declining number of firms actually introducing a new chemical entity in the United States. The declining importance of small firms in innovation will in due course lead to greater concentration of sales within larger firms, although total industry sales would never become entirely as concentrated as innovation due to the existence and growth of the generic drug sector. The differing data for the United Kingdom in Table 2-13 will be discussed later.

The deteriorating position of smaller firms has led to greater dependence by these firms on outside sources of innovation, licensed by the smaller firm for its own distribution with profits split by agreement between innovator and distributor. To demonstrate this dependence, 30 U.S.-owned, U.S.-located firms were segregated into three categories, each category accounting for about one-third of total U.S.-located pharmaceutical sales. The large-firm category contained those firms with the greatest sales volume (4 firms). The mid-size category contained 6 firms, and the small-firms category almost 20 firms. Table 2-14 presents trends on the origins of drugs marketed by these three size classes of firms. Examination of these trends demonstrates that, while the largest U.S.-located firms continue to self-originate (or develop in-house) the NCEs they market, smaller and even mid-size firms rely for a large and increasing proportion of introductions on licensed drugs.

This dependence of smaller and mid-size firms on licensed innovation renders these firms more vulnerable competitively for

TABLE 2-11 Worldwide Pharmaceutical Sales of Major Corporations, 1975 and 1980 (millions of dollars)

Firm	1975	1980
Hoechst	1,269	2,645
Bayer	939	2,371
Boehringer - Ingelheim	629	907
Schering AG	437	763
Germany	3,274	6,686
Roche/SAPAC	1,105	1,552
Ciba-Geigy	993	1,923
Sandoz Group	759	1,422
Switzerland	2,857	4,897
Merck & Co., Inc.	1,028	1,896
American Home Products	881	1,486
Bristol-Myers	768	1,357
Eli Lilly	690	1,156
Pfizer	683	1,291
Warner-Lambert	652	753
Abbott	614	1,379
Schering-Plough	484	852
Squibb	554	822
Upjohn	532	951
Johnson & Johnson	400	869
American Cyanamid	400	695
Dow	356	481[b]
Searle	319	590
Sterling Drug	299	702
Smith Kline	283	1,175
United States	8,943	16,455

Glaxo Group[a]	508	926
Wellcome Foundation[a]	492	996
Beechem Group[a]	385	705
ICI	289	805
United Kingdom	1,674	3,432
Takeda[a]	591	1,084
Japan		
Rhone-Poulenc	620	1,066
SANOFI	154	589
France	774	1,655
Akzo	308	665
Netherlands		
Montedison	301	529
Italy		

NOTE: Sales figures exclude intra-company transactions and include subsidiaries more than 50 percent owned. Sales have been converted to dollars using an exchange rate that consists of the average rate in the official exchange market during the fiscal year (calendar year unless otherwise shown). Some health products other than ethical drugs are included in sales in some cases.

[a]Fiscal year basis
[b]Prior to acquisitio of Merrell-National

SOURCE: Merck & Co., Inc., MSD Strategic Planning, 1981, West Point, PA. Most original data from corporate annual reports.

TABLE 2-12   Concentration of Sales in Domestic Pharmaceutical Markets, United States and United Kingdom

Concentration of Sales in the U.S. Ethical Drug Industry, 1957-1973

Year	4-Firm	8-Firm	20-Firm	N.E.
1958	28.8	50.9	79.5	24.27
1959	26.8	48.0	75.5	27.32
1960	25.8	47.3	75.4	28.25
1961	25.8	45.6	75.3	29.07
1962	25.4	44.3	74.5	29.76
1963	24.5	43.5	74.6	30.40
1964	23.7	42.2	74.1	31.06
1965	23.4	42.3	73.7	31.25
1966	24.4	42.7	74.1	31.15
1967	24.5	41.8	72.3	32.70
1968	25.4	43.6	74.4	30.86
1969	26.1	43.9	74.4	30.12
1970	26.3	43.2	73.6	30.77
1971	26.5	43.7	76.0	28.99
1972	27.6	43.6	75.4	28.90
1973	27.8	43.5	75.7	28.65

Concentration of Sales in U.K. Ethical Drug Market and Percentage of U.S. Market Accounted for by U.S. Firms, 1962-73

Year	4-Firm	8-Firm	20-Firm	N.E.	Share of U.S. Market Held by U.S. Firms
1962	29.9	46.8	80.7	24.63	46.9
1963	28.9	45.8	81.1	25.44	47.2
1964	27.9	44.7	79.6	26.95	45.9
1965	27.0	44.0	78.2	28.57	45.9
1966	26.3	42.9	76.7	28.65	45.2
1967	28.0	43.0	75.1	28.74	44.0
1968	29.7	44.4	75.1	27.78	42.8
1969	29.5	43.9	73.2	26.52	40.1
1970	29.7	44.1	73.2	28.65	39.4
1971	30.1	46.9	76.1	26.25	38.1
1972	29.1	45.9	75.2	27.22	38.6
1973	28.8	45.5	75.3	27.56	38.4

SOURCE: Henry Grabowski and John Vernon, "Structural Effects of Regulation on Innovation in the Ethical Drug Industry," in Robert Masson and David Qualls, eds. *Essays on Industrial Organization in Honor of Joe Bain*, Ballinger, Cambridge, 1976.

TABLE 2-13 Concentration of Innovation in Domestic Pharmaceutical Markets, United States and United Kingdom

Concentration of Innovational Output in the U.S. Ethical Drug Industry

Period	Total Number of New Chemical Entities (NCEs)	Number of Firms Having an NCE	Concentration Ratios of Innovational Output		
			4-Firm	8-Firm	20-Firm
1957-61	233	51	46.2	71.2	93.1
1962-66	93	34	54.6	78.9	97.6
1967-71	76	23	61.0	81.5	97.8

Innovational output is measured as new chemical entity sales during the first three full years after product introduction.

Data Sources: List of New Chemical Entities in each year obtained from Paul de Haen *Annual New Product Parade*, various issues; all information on ethical drug sales obtained from Intercontinental Medical Statistics.

Concentration of Innovational Output in the U.K. Ethical Drug Industry

Period	Total Number of New Chemical Entities (NCEs)	Number of Firms Having an NCE	Concentration Ratios of Innovational Output		
			4-Firm	8-Firm	20-Firm
1962-66	115	48	63.1	76.6	94.1
1967-71	95	44	42.7	66.4	91.1

Innovational output is measured as new chemical entity sales in U.K. during the first three full years after product introduction.

NOTE: Preliminary calculations suggest that innovative concentration in the United States has been declining in recent years in contrast to the apparent trend in this table which ends with 1976.

SOURCE: Henry Grabowski and John Vernon, "Structural Effects of Regulation on Innovation in the Ethical Drug Industry," in Robert Masson and David Qualls, eds. *Essays on Industrial Organization in Honor of Joe Bain*, Ballinger, Cambridge, 1976.

TABLE 2-14 Source of Approved NCEs Marketed by U.S.-Owned Firms Stratified by Firm-Size

	1963-1968	1969-1974	1975-1980
Self-Originated (%)			
Small	50.00	51.72	41.67
Middle-sized	55.56	26.67	52.63
Large	77.27	80.00	83.33
Acquired from U.S. Sources (%)			
Small	20.59	17.24	22.22
Middle-sized	0.00	26.67	21.05
Large	4.55	0.00	0.00
Acquired from Foreign Sources (%)			
Small	29.41	31.03	36.11
Middle-sized	44.44	46.67	26.32
Large	18.18	20.00	16.67

Source of INDs Filed by U.S.-Owned Firms Stratified by Firm-Size

	1963-1968	1969-1974	1975-1980
Self-Originated (%)			
Small	74.19	69.72	56.00
Middle-sized	74.24	85.71	78.57
Large	96.39	93.67	89.74
Acquired from U.S. Sources (%)			
Small	7.26	3.67	4.00
Middle-sized	4.54	4.40	1.79
Large	2.41	0.00	0.00
Acquired from Foreign Sources (%)			
Small	18.55	26.61	40.00
Middle-sized	21.21	9.89	19.64
Large	1.20	6.33	10.26

NOTE: The ranking of firms is based on their total domestic U.S. pharmaceutical sales for 1977-1978 as described and listed in the 1978 edition of the *Medical and Healthcare Marketplace Guide.*

SOURCE: Center for the Study of Drug Development, University of Rochester.

two reasons. First, lower rates of return may be incurred on licensed drugs due to license fees that must be paid to the innovator. This smaller cash flow provides less funding for internal research to self-originate new drugs. Second, the substantial reliance on foreign firms for licensed NCEs is based on the fact that few foreign-owned firms (other than the Swiss) have until recently established extensive subsidiaries in the United States. Should this arrangement be altered by establishment of such U.S.-located subsidiaries, a substantial source of sales for U.S.-owned, U.S.-located firms could gradually disappear. In fact, numerous foreign-owned firms have indeed entered the U.S. market in recent years, often by purchasing smaller U.S. firms. Table 2-15 indicates the extent of this multinational diffusion into the United States. It is interesting to note that almost all of the indicated entry has derived from European-owned firms, suggesting that should Japanese-owned firms later attempt such entry, fewer appropriate small firms will be available for similar entry by merger or purchase.

It is in fact the increased multinational diffusion of nationally owned pharmaceutical firms which makes it difficult to interpret the British concentration data previously given in Table 2-13. The entry of foreign-owned subsidiaries into the British market could offset any contraction in innovation or sales by smaller British-owned firms. Nonetheless, it is interesting to know that in a period when U.S.-located innovation became concentrated into fewer firms, no such concentration was observed in certain foreign markets. Indeed, roughly during this period, the West German-located market became slightly more competitive, with the market share of the five leading firms falling from 27 percent in 1975 to 26 percent in 1979, and more significantly the share of the top ten firms falling from 45 to 40 percent during the same time period.[10]

TRADE

Pharmaceutical products have traditionally provided a surplus for the U.S. trade balance (see Table 2-16). Yet, this surplus in absolute terms is not significantly greater than that of Switzerland, West Germany, or the United Kingdom, despite the substantially larger level of U.S. production. This imbalance arises because the U.S. exports a much smaller fraction of production than do its prime competitors (as shown in Table 2-17). This lower level of exports as a proportion of domestic production provides the United States with a currently roughly equivalent share of world pharmaceutical exports (see Table 2-18), a share that has markedly deteriorated since 1950. In part, this low

TABLE 2-15 Foreign-Based Participants in the U.S. Pharmaceutical Marketplace

Foreign Company	U.S. Subsidiary	Year of Entry
Long-standing Participants		
AB Astra	Astra Pharmaceuticals	
Hoffmann-La Roche	Roche Labs	
The Wellcome Foundation	Burroughs-Wellcome	
Sandoz, Ltd.	Sandoz Pharmaceuticals	
	Dorsey Labs	
Akzona, Inc. (AKZO)	Organon Pharmaceuticals	
Ciba-Geigy AG	Ciba	
	Geigy	
	Alza	1977
	S. J. Tutag	1979
Recent Entrants		
Hoechst AG	Hoechst-Roussel	about 1966
	Calbiochem Corp.	1977
Bayer AG	Cutter Labs	1973
	Miles Labs	1978
	Dome Labs	1978
C. H. Boehringer Sohn	Boehringer Ingelheim Ltd.	1973
	Hexagon Labs	1975
	Philips Roxane	1979
Fisons, Ltd.	Fisons Corp.	1973
ICI Ltd.	Stuart Pharmaceuticals	about 1968
Beecham, Inc.	Beecham Labs	1969
Byk-Gulden, Inc.	Savage Labs	?
Montedison	Adria Labs (50%)	1974
	Warren-Teed	1977
Boots, Ltd.	Rucker Pharmacal	1977
Glaxo Group, Ltd.	Meyer Labs	1977
Nestle Alimentana SA	Alcon Labs	1977
	Lafayette Pharmacal	1978
	Burton Parsons	1979
SANOFI SA	Towne Paulsen	1975
	Generic Pharm. Corp.	1976
	Western Research Labs	1976
Connaught Labs, Ltd.	Swiftwater Biological Unit	1978
Rhone-Poulenc, SA	Norwich-Eaton (10.5%)	1978
Green Cross Corp.	Alpha Therapeutics	1978
Mitsubishi Chem. Ind.	Key Pharmaceuticals (10%)	1979
Schering AG	Berlex Labs	1979
Kali-Chemie AG	Purepac Labs	1979

SOURCE: Merck & Co., Inc., MSD Strategic Planning, West Point, PA.

proportion of production devoted to exports is associated with the relatively more extensive multinational scope of U.S.-owned firms, and their reliance on sales rather than exports. Equally important is the traditional relative unimportance of exports to U.S. producers, as may be seen by comparison of total U.S. exports to Gross National Product (GNP). From this perspective, the U.S. pharmaceutical industry is typical of other sectors of the American economy. Only the United Kingdom and Switzerland

TABLE 2-16 Balance of Pharmaceutical Trade, Current Account, Selected Nations (millions of dollars)

	1965	1970	1975
West Germany	164	316	528
United States	198	335	639
United Kingdom	156	254	611
Switzerland	147	251	669
France	55	86	293
Italy	2	11	40
Japan	-27	-150	-316

SOURCE: United Nations, *Yearbook of International Trade Statistics*, UN, New York, various years.

export a significantly higher share of pharmaceutical production than of total GNP. An interesting feature of international pharmaceutical markets is the relative isolation of the United States and especially Japan from direct international trade in ethical drugs (see Tables 2-17 and 2-19).

SUMMARY

A basic conclusion of any overview of the preceding data is that the competitive position of the U.S. pharmaceutical industry has

TABLE 2-17 Exports as a Proportion of Domestic Production, Selected Nations (percentages)

	1965		1970		1975	
	Pharm.	GNP	Pharm.	GNP	Pharm.	GNP
West Germany	25	19	28	22	24	26
United States	6	5	6	5	11	8
United Kingdom	27	18	45	22	55	26
Switzerland	90	30	91	35	na	30
France	11	14	18	16	21	18
Italy	11	17	17	19	17	21
Japan	na	11	2	11	2	14

NOTE: Pharmaceutical figures give the ratio of pharmaceutical exports to domestic pharmaceutical production, while GNP figures give the ratio of dollar volume of all exports to dollar value of GNP.

SOURCES: Organization for Economic Cooperation and Development, *The Chemical Industry*, OECD, Paris, various years.

United Nations, *Yearbook of International Trade Statistics*, UN, New York, various years.

International Monetary Fund, *International Financial Statistics*, VXXXIII, No. 8, August 1980.

TABLE 2-18 Exports of Pharmaceuticals, Selected Nations Market Share (percentages)

	1955	1960	1965	1970	1975
West Germany	10	12	16	19	16
United States	34	30	16	15	13
United Kingdom	16	14	13	13	13
Switzerland	14	13	14	13	13
France	12	11	11	9	10
Netherlands	3	5	5	6	5
Italy	3	4	5	6	6
Japan	1	2	3	2	2

SOURCE: United Nations, *Yearbook of International Trade Statistics*, UN, New York, various years.

deteriorated, especially in the earliest phases of the discovery/development marketing process. The trend of this decline has not been constant, but rather has proceeded rapidly in the early 1960s, followed by more gradual movement. The initial decline occurred roughly in the years 1962 to 1968. During this period, the U.S. share of world pharmaceutical exports was halved, market shares for sales deteriorated markedly, the number of firms producing NCEs was halved, and the U.S. share of nationally located R&D dropped significantly. Subsequently there has been rough stability in terms of shares of innovation, exports, and national market sales and of number of innovating firms, while U.S. shares of production, patent, clinical trials, and research have exhibited continued gradual decline.

TABLE 2-19 Pharmaceutical Imports as a Proportion of Domestic Consumption, Selected Nations (percentages)

	1965	1970	1975
West Germany	8	12	13
United States	2	2	3
United Kingdom	6	16	24
Switzerland	64	70	na
France	7	12	13
Italy	11	16	16
Japan	na	7	7

NOTE: Apparent consumption is computed as the sum of production and imports minus exports.

SOURCES: Organization for Economic Cooperation and Development, *The Chemical Industry*, OECD, Paris, various years.

United Nations, *Yearbook of International Trade Statistics*, UN, New York, various years.

For the near future, there are numerous indications that recent stability in U.S. shares of introduction and sales may not persist. Declining U.S. shares of R&D expenditures and NCEs beginning clinical trials should, in time, lead to falling U.S. shares of introductions and sales, though the exact magnitudes and timing of these downturns are impossible to predict.

By way of conclusion, it should once again be stressed that for the foreseeable future U.S. pharmaceutical firms will remain innovative and growing. Available data simply indicate foreign firms in this industry will be even more innovative and will grow even more rapidly. The result is a diminished though still vital U.S. presence, one of the more significant of high-technology industries.

NOTES

1. Charles River Associates, op. cit., p. 56.

2. Figures by the U.S. Pharmaceutical Manufacturers' Association, cited in Barrie James, The Future of the Multi-national Pharmaceutical Industry to 1990, New York, John Wiley, 1977, p. 71.

3. William M. Wardell, et al., "Development of New Drugs Originated and Acquired by United States-Owned Pharmaceutical Firms, 1963-1976," Clinical Pharmacology and Therapeutics, Vol. 28, no. 2.

4. U.S. Pharmaceutical Manufacturers' Association, Factbook, 1980, PMA, Washington, D.C., 1981.

5. See Peter Barton Hutt, "The Importance of Patent Term Restoration to Pharmaceutical Innovation," Health Affairs, Vol. 1, No. 2, Spring 1982. and especially Ronald Hansen "The Pharmaceutical Development Process: Estimates of Development Costs and Times and the Effects of Prepared Regulatory Changes," in Robert Chein, ed., Issues in Pharmaceutical Economics, D.C. Heath & Co., 1979. Hansen's original estimate was $54 million in 1976 dollars, which converts to $70 million in 1980 dollars.

6. A recent study at the Center for the Study of Drug Development shows that the number of compounds entering clinical testing between 1958 and 1962 was more than double the number entering clinical testing between 1963 and 1979. These data are based on approximately 50 percent of all NCE research by U.S.-located pharmaceutical firms. See M. S. May and W. M. Wardell, New Drug Development During and After a Period of Regulatory Change: Research Activity of Major United States Pharmaceutical Firms, 1958 to 1979.

7. Estimates of the higher average cost of domestic U.S. pharmaceutical R&D have been provided by Lewis Sarett, "FDA

Regulations and Their Influence on Future R&D," _Research Management_, March 1974.

8. For data on absolute production levels, see _The Chemical Industry_, OECD, Paris, annual issues.

9. For a study of returns to scale in pharmaceutical innovation, see Henry Grabowski and John Vernon "Structural Effects of Regulation on Innovation in the Ethical Drug Industry," in Robert Masson and David Qualls, eds., _Essays on Industrial Organization in Honor of Joe Bain_, Bullinger, Cambridge, MA., 1976.

10. IMS, _Pharmaceutical Marketletter_, September 22, 1980.

3
Determinants of National Pharmaceutical Competitive Advantage

Essentially, four broad hypotheses may be presented as explanations for the recorded shifts in competitive position of U.S. pharmaceutical firms.

* Microeconomic factors--basic conditions of cost, input supply, and output demand have operated adversely for U.S. pharmaceutical firms.
* Macroeconomic factors--the relative decline of the U.S. pharmaceutical industry is simply part of an economy-wide deterioration of American industrial position.
* Regulatory factors--the comparatively more costly and extensive regulations of the U.S. Food and Drug Administration (FDA) have disadvantaged U.S. firms.
* Artificial economic supports and restraints--the discriminatory tax and trade policies of foreign governments have unfairly advantaged foreign-located firms.

Of these hypotheses, the first would appear to be of little explanatory value, while the second provides an important but only partial accounting for the documented relative decline. The sequence of hypotheses considered below begins with three microeconomic aspects (labor costs, domestic growth of demand, and national supply of qualified scientists), next introduces macroeconomic, then regulatory aspects, and concludes with two discriminatory policy aspects (taxation and trade barriers).

LABOR COSTS

Comparative levels of wages and salaries are a generally important determinant of industrial location. The relative levels of compensation for both manufacturing and research staff among North America, western Europe, and Japan are thus potentially

53

TABLE 3-1   Manufacturing Wages (1970=100)

	Hourly Earnings in Manufacturing Gross Earnings per Production Worker		Hourly Rates in Manufacturing				Monthly Earnings in Manufacturing
	U.S.	West Germany	U.K.	France	Italy	Switzerland	Japan
1960	67	44	57.0	44.9	41.2	60	31.1
1965	78	69	71.8	64.4	67.5	78	50.3
1970	100	100	100.0	100.0	100.0	100	100.0
1975	143	160	220.1	188.3	241.4	155	230.3
1980	216	216	439.9	374.4	631.7	180	345.2

SOURCE: "Main Economic Indicators," OECD, Paris, various years.

important sources of explanation for differential patterns of growth in research and production. Data on trends of general national compensation in manufacturing are reported in domestic currencies in Table 3-1. While rates of growth of wages in domestic currencies have varied widely during the last two decades, these domestic trends are mostly offset by opposite movements in exchange rates. Thus, while Swiss manufacturing compensation has risen more slowly than that in the United States, the Swiss currency has (until quite recently) persistently appreciated against the dollar. As a consequence, the dollar cost of Swiss labor relative to U.S. labor has moderately increased. Thus, this first potential explanation of the deterioration of the U.S. competitive position fails to account for the observed decline, as U.S. wages have not risen more rapidly than foreign costs.

MARKET GROWTH

Consumer demands for pharmaceutical products have increased in all nations, but at widely divergent rates (see Table 3-2). Data on consumption levels indicate that foreign markets have uniformly grown more rapidly than those of the United States. However, the parallel, extensive growth of foreign production is not completely explained by these figures as they fail to indicate the reason this growth was met by production abroad rather than by U.S. exports.

TABLE 3-2 Domestic Pharmaceutical Sales, Selected Countries, Selected
Years (millions of dollars)

	1965	1970	1975	Growth
United States	3,121	4,701	7,387	9
Japan	1,298	2,975	6,402	17
West Germany	742	1,408	3,952	18
France	967	1,207	2,731	11
Italy	514	920	2,181	16
Spain	236	597	1,652	21
United Kingdom	300	408	815	10

NOTES: Growth figure is annual percentage growth rate. Sales are apparent consumption (production plus imports minus exports) except for US and UK.

SOURCES: US PMA, *Prescription Drug Industry Factbook*, PMA, Washington, D.C., 1976.

ABPI, *Annual Report*, London, various years.

NATIONAL SCIENTIFIC CAPACITY

The general research base of a nation would be expected to affect strongly the extent and success of industrial R&D. One aspect of this research base is illustrated in Table 3-3, where the U.S. proportions of articles published in various fields are shown. Such American accomplishments not only fail to provide an explanation for the falling U.S. share of industrial pharmaceutical research and innovation, but suggest that no such trends should exist.

A crucial aspect of the national scientific base is governmental and nonprofit expenditures on research, which in the United States amounts to about 75 percent of all health-related R&D. Data on these expenditures are presented in Table 3-4 and demonstrate the stunningly large proportion of world health expenditures that the U.S. government funds. Indeed, it would appear (based on plausible extrapolation from Table 3-4) that the U.S. government expends, by itself, virtually as much for health research as do all other sources, both industrial and governmental, in the western developed nations.

GENERAL RELATIVE DECLINE OF U.S. INDUSTRY

The relative decline of the U.S. pharmaceutical industry is unfortunately not unique. Numerous American industries, if not our entire economy, have exhibited sustained decay of relative position over the past two decades. Several reasons exist for the relatively more vibrant growth of foreign economies. Much of

TABLE 3-3   U.S. Proportion of the World's Articles, 1973-1977 (percentages)

Field	1973	1975	1977
All	39	38	38
Clinical medicine	43	43	43
Biomedicine	39	39	39
Chemistry	23	22	22
Biology	46	45	42

NOTE: Articles are counted from the 271,000 to 279,000 articles, notes, and reviews per year from over 2,100 of the influential journals of the *Science Citation Index*, Corporate Tapes of the Institute for Scientific Information.

SOURCE: National Science Board, Science Indicators 1976, National Science Foundation, Washington, D.C., 1977.

European and Japanese manufacturing capacity was devastated by World War II, and postwar recovery of these economies brought on inevitable correction to the early 1950s economic dominance of the United States. Secondly, particularly in the case of Japan, standards of living abroad were historically below those of the United States, and the international diffusion of manufacturing technology brought a leveling of national productivities that implied more rapid growth abroad. Finally, differences in national industrial policies and management procedures contributed to the differential national economic records. By the mid-1970s, three decades after World War II, the latter policy differences were probably preeminent in significance.

Tables 3-5 and 3-6 indicate that decay in U.S. shares of sales and innovation has occured for many industries, not just pharmaceuticals. These general relative declines suggest that if deterioration of competitive position in the pharmaceutical industry is no worse nor no better than that of most industries, especially most high-technology industries, then there is no need for arguments unique to the pharmaceutical industry to explain these firms' relative decline. Instead, contemporary economy-wide factors such as taxation, investment policy, export policy, national levels of savings and investment, and so on must be invoked as hypotheses. From this viewpoint, Tables 3-5 and 3-6 would suggest precisely that no such additional industry-specific explanations are needed in the case of the pharmaceutical industry.

TABLE 3-4 Sources of Expenditures for Health Research and Development, Selected Nations (millions of dollars)

	Year	Pharmaceutical Industry	University	Government
France	1969	63	7	na
West Germany	1972	310	na	168
Switzerland	1975	244	17	na
United Kingdom	1972	108	3	200
United States	1972	535	50	2,223

SOURCE: Organization for Economic Cooperation and Development, *Impact of Multinational Enterprises on National Scientific and Technical Capacity: Pharmaceutical Industry*, O.E.C.D., Paris, 1977.

Figure 3-1 and Table 3-7 suggest otherwise. Figure 3-1 demonstrates that the pharmaceutical industry has behaved uniquely among the west European and Japanese chemical industries, while in the United States, the performance of the pharmaceutical industry is literally indistinguishable from other components of the general chemical industry. Something unique indeed has occurred. More significantly, Table 3-7, when compared with Table 2-1, suggests an even more profound difference. The U.S. share of world chemical R&D has fallen only moderately during roughly the same time period when the U.S. share of pharmaceutical R&D has drastically dropped. Indeed, none of the industries subject to the OECD study that underlies Table 3-7 exhibits a fall or research share that in any way parallels the severity of the drop in the pharmaceutical industry. For an explanation of those factors that have so severely affected relative U.S. pharmaceutical R&D, attention must be turned elsewhere.

INDUSTRIAL POLICY: REGULATION

The costliness and success of pharmaceutical innovations are vitally affected by regulation of drug safety and effectiveness. The detail and pervasiveness of this regulation in the United States are almost unique, both in comparison with U.S. regulation of other industries and with foreign regulation of pharmaceutical markets. Further, the scope and volume of U.S. pharmaceutical regulations have dramatically expanded since 1960, suggesting that changes in this component of U.S. industrial policy may well have provided substantial effect on the competitive position of U.S. firms.

TABLE 3-5   Sales of Major U.S. Corporations as a
Proportion of Major World Corporations, Selected
Industries, 1959 and 1978 (percentages)

	1959	1978
High Technology Industries		
Aerospace	95.4	90.1
Chemicals	66.3	31.9
Electronics-appliances	75.6	46.9
Pharmaceuticals	61.1	35.0
Other Industries		
Automotive	84.3	59.7
Food products	66.6	55.7
General machinery	61.7	51.8
Metal manufacturing	89.9	32.4
Metal products	66.8	43.2
Paper and paper products	92.2	70.6

SOURCE: Nestor Terleckyj, "Technology and the Changing Position of U.S. Firms Among the World's Largest Companies," paper
given at the December 1979 meeting of the New York State Bar
Association, Antitrust Law Section.

Original data from *Fortune* (July, August 1960 and May, July,
August 1979).

Legislative foundations for regulation of ethical drugs in the
United States are the Food and Drugs Act (1906), the Federal
Food, Drug, and Cosmetic Act (1938), and the (Kefauver-Harris)
Drug Amendments (1962). The essence of this regulation, since
1938, has been premarket approval by the U.S. Food and Drug
Administration (FDA) for any new drug product. In other words,
commerce for a new drug is prohibited until it is adequately tested
for safety and, since 1962, for effectiveness in treatment for
indications prescribed on its label. Exemptions to this requirement are allowed only for investigational use by qualified scientific experts. The basic mechanics of this requirement involve
submission of data in the form of an NDA.

Before 1962, unless the FDA acted to reject the NDA within
90 days of submission, the new drug could be marketed. Important changes in FDA requirements emerged after adoption of the
1962 Amendments. Central provisions of the Amendments are:

• Effectiveness must be demonstrated by the manufacturer
through "adequate and well-controlled investigations" to obtain
FDA approval of an NDA.
• FDA monitors investigational drug studies in humans by
requiring data in the form of a Notice of Claimed Investigational
Exemption (IND). If FDA vetoes the IND, testing in humans may
not begin.

TABLE 3-6   Patents Granted to U.S. Applicants as Proportion of Total
U.S. Patents, Selected Industries, 1963 and 1973 (percentages)

	1963	1973	Change
All industries	80.5	68.2	–12.3
Aerospace	74.9	58.8	–16.1
Chemicals	74.3	63.1	–11.2
Electrical	83.6	69.8	–13.8
Pharmaceuticals	64.0	54.7	–9.3

SOURCE: Nestor Terleckyj, "Technology and the Changing Position of U.S.
Firms Among the World's Largest Companies," paper given at the December 1979
meeting of the New York State Bar Association, Antitrust Law Section.

• FDA must affirmatively approve an NDA, rather than allowing automatic approval after 90 days.
• FDA must establish good manufacturing practice regulations.

Many scholars have argued that the 1962 Amendments were responsible for greatly increasing FDA regulation. Following the 1962 Amendments, the scope and intensity of U.S. pharmaceutical regulation significantly increased. A closer examination suggests that the specific requirements of the 1962 Amendments do not fully account for all post-1962 changes in FDA regulation. Some of the contemporary structure of regulation for new drugs might have emerged without congressional adoption of the Amendments. In a recent essay, David Weimar has postulated how the IND procedure might have developed:

Under Section 505(i) of the 1938 Law, the FDA had authority to promulgate regulations that governed the distribution of drugs for investigational use. The regulations initially issued by the FDA required manufacturers to keep records of the distribution of drugs for investigational purposes, investigators to sign statements that they had adequate training and facilities to safely conduct the investigations, and labels to contain the statement, "Caution: New Drug--Limited by Federal Law to investigational use." These regulations permitted the distribution of thalidomide to over 1200 physicians. The same furor that led to passage of the 1962 Amendments prompted the FDA to revise its regulations for investigational drugs. In August of 1962, prior to passage of the Kefauver-Harris Amendments, the FDA published regulations that in effect established an IND procedure. They required that only

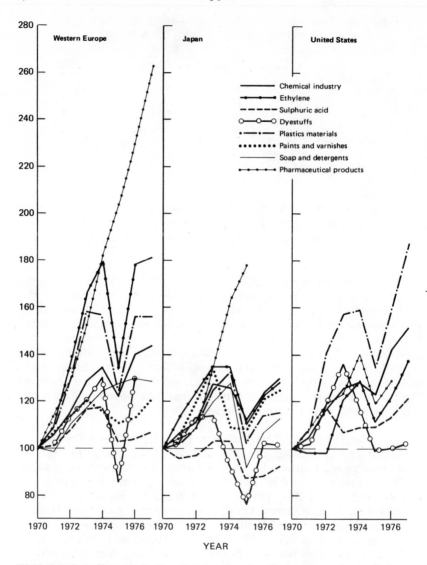

FIGURE 3-1 Trends in Production in the Major Branches of the Chemical Industry

SOURCE: OECD, The Chemical Industry, Paris, 1978.

TABLE 3-7   U.S. Share of Industrial R&D Expenditures in Nine
OECD Countries, 1967 and 1975 (percentages)

	1967	1975	Change
Aerospace	81.5	72.6	−8.9
Electrical	66.3	57.4	−8.9
Chemical	44.5	40.2	−4.3
Other transport	58.7	48.7	−10.0
Machinery	55.2	53.2	−2.0
Basic metals	37.8	40.1	+2.3
Chemical-linked	43.0	43.1	+0.1
Other manufacturing	49.7	49.8	+0.1

NOTE: Countries surveyed include U.S., Japan, West Germany, France, U.K., Belgium, Italy, Canada, and Sweden.

SOURCE: Organization for Economic Cooperation and Development, *Trends in Industrial R&D in Selected OECD Member Countries, 1967-1975*, OECD, Paris, 1979.

qualified investigators be used; that their qualifications be filed with the FDA; that drugs be tested on humans only after animal testing; that the FDA be kept fully informed of the results of the testing; and that special precautions be taken in the testing of drugs intended for use by children or pregnant women. This is one of several instances we will encounter where proposed legislative changes have been anticipated in FDA regulations.[1]

Nonetheless, it is clear that the 1962 Amendments--and particularly the new requirement of affirmative FDA approval of an NDA--had a major impact on regulatory requirements for new drugs.

An uninterrupted series of requirements (some explicitly required by the 1962 Amendments and others adopted as adjuncts to those Amendments) have been issued since that year, a partial listing of which follow:

1963   Regulations specify good manufacturing practice.
1966   Preclinical guidelines issued for reproductive, teratology, and perinatal and postnatal studies.
1968   Preclinical guidelines issued for toxicity testing.
1970   Regulations specify requirements for "well-controlled investigations" to produce "substantial evidence" of efficacy.
1970   30-day delay for initiation of testing in humans after submission of IND.

1972   Preclinical guidelines issued for chemistry, expanding requirements for drug manufacture, and quality control.
1975   Freedom of Information Act regulations issued.
1977   Clinical guidelines issued for various drug classes.
1978   Regulations specify Good Laboratory Practices. These regulations issue standards for test protocols, quality control, recordkeeping, equipment, buildings, and facilities, etc.

The impact of the 1962 Amendments was heightened by two factors. A first factor has been scientific advancement in medical technology that enables more acute detection of potential adverse reactions. As the ability of researchers to test for safety and efficacy improved, the expectations and requirements of the FDA increased for the volume and quality of premarket testing.

The second and most critical element, however, has been intense political pressure on the FDA. The very nature of the premarket approval system confronts the FDA with difficult choices. The limited clinical trials used statistically to ascertain safety and effectiveness of new drugs cannot possibly provide, with certainty, estimates of the extent of therapeutic benefits and of adverse reactions. Any drug potent enough to be effective carries some risk of adverse reaction in humans and adverse results in test animals. It is widely known, for example, that penicillin has fatal effects on guinea pigs and aspirin has teratological effects on animals (suggesting that either product would today encounter severe regulatory obstacles for approval by the FDA). Under these circumstances, the FDA must weigh patient risk of adverse reactions against patient risk of disease due to inferior or no pharmaceutical treatment.

This complex calculus must inform FDA decisions as to what tests, guidelines, and requirements should be enforced on pharmaceutical firms seeking clearance for new products. Yet, the American political process places substantial pressure on the FDA not to approve new drugs. The Congressional Information Service Index for 1969 and 1970 reports congressional hearings on FDA decisions concerning 38 specific drugs or drug classes. Of these hearings, only two questioned FDA decisions not to approve marketing of drugs; one of the latter hearings was on laetrile. Former FDA Commissioner Schmidt, in 1974, summarized the impact of this political pressure as follows:

> In all our history, we are unable to find one instance where Congressional hearings investigated the failure of FDA to approve a new drug....[T]he message conveyed by this situation could not be clearer....Until perspective is brought to the legislative oversight function, the pressure from Congress for FDA to disapprove new drugs will continue to be felt, and could be a major factor in health care in this country.[2]

The process of regulation in other nations differs significantly in several respects from that in the United States, generally being more flexible and prompt. A recent U.S. General Accounting Office (GAO) report has identified several key distinctions in operation of pharmaceutical regulation abroad, some of the more important of which are excerpted below.

Greater Use of Expert Committees

In most European nations, decisions on approval of new drugs are not solely the responsibility of career bureau officials; but instead, these decisions are either substantially advised or formally made by committees of independent medical experts.

According to European regulatory and industry officials, using a committee of experts insulates the regulatory authority from public criticism, gives credence to the final decision, and expedites the review and approval of drugs.

Some European committees of experts are mandated to review all drug applications and either approve a drug when it is shown to be safe and efficacious or recommend to the regulatory agency that a drug should or should not be approved. In three countries--the Netherlands, Norway, and Sweden--the committees had been given the responsibility to make the decision to approve, reject, or withdraw a drug. The United Kingdom's committee only advises the government agency on the safety and efficacy of a drug; however, we were told that its recommendations have always been followed.

At FDA, committees are used to provide advice on problems or questions FDA may have concerning selected drug applications. However, applications are not submitted routinely to the committees in the United States as they are in foreign countries. FDA has sole responsibility for making a decision on an application based on the scientific data submitted and any advice from the expert committee.

Greater Acceptance of Foreign Data

The traditional refusal by FDA to accept foreign data as a basis for NDA approval has required costly delay and new testing.

Foreign clinical study data are accepted by most foreign drug regulatory agencies as evidence of a drug's safety and efficacy if the studies are well-conceived, well-controlled, performed by qualified experts, and conducted in accordance with acceptable ethical principles. Domestic verification is sometimes required. According to foreign government officials, the degree of additional domestic verification depends on such factors as the source of the original clinical trials, since medical practices and hereditary, dietary, and other factors may be different from those of the registering country. Some countries--the Netherlands, Norway, and Switzerland--accept foreign data submitted without domestic verification depending on the source. Other countries--Sweden, and the United Kingdom--will normally request some domestic verification.

Although the FDA may have accepted, in some cases, foreign data as pivotal evidence of the safety and efficacy of a drug, its policy in this regard is not clear. Officials of the drug firms we visited, indicated that FDA would not accept foreign data as primary pivotal evidence, and required that the safety and efficacy of a drug be supported on the basis of duplicate domestic studies. FDA's Director of the Bureau of Drugs stated that FDA has had a reputation for not accepting foreign data. We believe FDA needs to formally clarify and communicate its policy on the acceptance of foreign data.

Less Politicization of Drug Approval Process

In the European countries we visited, drug regulatory officials told us there was no direct parliamentary or consumer scrutiny on the drug regulatory process. When a parliamentary body wishes to inquire about issues concerning drug regulatory policies, procedures, or decisions, drug regulatory officials are not required to appear before the parliament and thus are not subjected to parliamentary pressures. Rarely, if ever, is the regulatory agency's director or any of its employees asked to appear before the parliament. Instead, the minister of health, who is a member of the parliament, responds to inquiries from parliament on drug regulatory matters.

Foreign drug regulatory officials advised us that members of parliament in their countries, for the most part, believe that the regulatory agency has primary responsibility for regulating drugs and that parliamentary involvement should be minimal.

FDA's drug regulatory process comes under intensive congressional oversight and scrutiny by consumer-oriented organizations. Officials of many U.S. drug firms told us that congressional and consumer scrutiny tends to slow FDA's drug approval process.

Greater Cooperation between Regulators and Industry

Most foreign drug industry officials explained that they have easy access to British, West German, Swiss, Norwegian, and Swedish experts and drug regulatory officials for frequent and open scientific discussions off the record. According to these officials, scientific discussions address the tests necessary for approval and other difficulties, and in their opinions assist in developing a framework for clinical trials.

American drug firm officials told us that FDA appears to favor an adversary relationship with industry. Bureau of Drug reviewers, according to these officials, review an application with the attitude that there are errors in the application and that they must find them. This adversary attitude is compounded by a communications problem between FDA and industry. According to drug firm officials, FDA has become increasingly inaccessible. One drug firm official told us "Industry is becoming more isolated from FDA. Bureau of Drug reviewers will not use phones to ask us questions they have on an NDA." Another drug firm official, in comparing FDA reviewers with their European counterparts said, "Medical officers are a lot more open and frank in Europe. As a result, they are able to resolve problems with NDA submissions in a more timely manner in Europe."[3]

A clearly demonstrable effect of the totality of differences between FDA regulation and that abroad is longer approval times in the United States as compared with most other nations. The above cited GAO report indicated that the mean times between application for marketing new drugs and regulatory approval of applications are as follows (for selected nations):

Canada	16 months
Norway	17 months
Sweden	28 months
United Kingdom	5 months
United States	23 months

TABLE 3-8 Comparison of U.S. NCE Introduction Dates with UK, France, and West Germany (all U.S. NCE introductions between 1963-1967)

Country and Year of U.S. Introduction	Number (Percent) Introduced			
	Before U.S.	Same Year	After U.S.	Not Abroad
United Kingdom				
1963-1967	30 (48)	12 (19)	21 (33)	13
1968-1975	44 (61)	14 (19)	14 (19)	26
West Germany				
1963-1967	21 (46)	12 (26)	13 (28)	28
1968-1975	39 (56)	17 (24)	14 (20)	30
France				
1963-1967	11 (27)	4 (10)	26 (63)	31
1968-1975	26 (45)	9 (16)	23 (40)	43

SOURCE: Henry Grabowski, "Regulation, The Innovative Process, and International Diffusion in the Pharmaceutical Industry," mimeograph, 1979.

The GAO report further identifies several important drugs that have been introduced abroad significantly earlier than in the United States.

An inevitable effect of FDA delay in approval of new drugs is earlier introductions of new pharmaceutical products abroad. The United Kingdom and West Germany now receive NCEs earlier than does the United States, and the U.S. position has moved from one of lead to lag as regards the diffusion of pharmaceutical innovation, as can be seen in Table 3-8.

An additional impact of increased U.S. regulation of the pharmaceutical industry has been increased costs of development for NCEs. As the recent study for OTA by Leonard Schriffin explains:

> Regulations of the sort contained in the 1962 Amendments raise a firm's costs of drug development by requiring inputs into the R&D process, reduce its chances of R&D success, and delay the time of pay-off for successful innovation. Economic theory tells us what further to expect from such cost increases. For one thing, they will alter the amount of R&D activity: firms finding it commercially infeasible to attempt to innovate will find that to be even more the case; those finding it marginally profitable to do so, may well find it now to be unprofitable; and firms that are active innovators will find that fewer of the available projects will remain advantageous to pursue. Thus, while the total

dollar volume of R&D outlays may be increased, there will be fewer inputs and outputs than otherwise associated with R&D activity.

The evidence is quite clear that, although R&D costs were rising prior to 1962, the Amendments accelerated the trend. These cost increases have influenced firm strategies discernibly, if unevenly, and the overall [U.S.] rate of innovation has been reduced as a result.[4]

The relatively early assessment of new drug candidates in the clinic is a particular advantage available to firms conducting research abroad. Probing studies of efficacy in patients must now be preceded by long and costly laboratory studies. The clinical usefulness of cortisone and even penicillin would have been seriously delayed under today's regulations in that they would have required the production of at least 20 to 30 Kg of cortisone before it would be allowed in the clinic. With the process as it stood from the first synthesis of about 10 to 15 grams of cortisone per 1000 pounds of desoxycholic acid, no company could have afforded to undertake the herculean task of meeting today's requirements.

Another aspect of the same situation is the selection of the best candidate for clinical studies. Today, it is a very sizeable task to take several related compounds to the clinic in order to determine which one would be the best to develop, a procedure that can more easily be done abroad. While American companies may be allowed the same opportunities to test several related compounds in patients at the same time, it would be very difficult to do so because the R&D expertise and organization knowledgeable about the drug candidates are located in the United States. This situation places American companies at a disadvantage.

INDUSTRIAL POLICY: TAXATION

Specific tax policies that benefit the pharmaceutical industry are not common among the major OECD nations. The most prominent examples of such policies occur in Japan. In the first place, Japan allows special tax treatment for industrial research and development expenditures. The relevant regulation is Tax Special Treatment Article 42-3. It states that a corporation can enjoy a tax credit if its net research expense spent and tax deductible in the current fiscal year (i.e., gross expense minus any subsidies from affiliated companies, government and others) exceeds the expenses of any previous fiscal year since 1967. In that event, 20 percent of the excess (i.e., current research expense minus the maximum expense of any year since 1967) is deducted from corporate tax for the year. A simple example would be:

Company Research Expense	$1,000
Government Research Grant	100
Net Research Expense	900
Highest Research Expense of any year since 1967	700
Difference	200

20% of 200 = 40

Total Corporate Tax for Year (say)	1,500
Less 20% of 200	40
Net Tax	1,460

Further, Japan maintains not one, but two agencies for publicly funding pharmaceutical research. The first is the Research Development Corporation of Japan, which finances technological work in all industries. Its most recent fundings have been:

1980	Green Cross Co.	$4.2 Million
1978	Kaken Yakkako Co.	3.8 Million
1977	Teijin Co.	2.9 Million

The second agency is the Council for New Drug Development Promotion that funds only pharmaceutical research. Recent Council expenditures have been:

1980	Kyoto University	$0.8 Million
1979	Takeda Chemical Co.	0.9 Million

These findings make up a pitifully small proportion of Japanese national expenditures on pharmaceutical research and development. Yet, because these public grants are directed toward basic research, their impact is presumably larger than simple percentages of total expenditures would indicate.

INDUSTRIAL POLICY: TRADE

Tariffs are of little consequence to the pharmaceutical industry. Rates are usually low and do not generally seriously affect trade. Non-tariff barriers are of much greater importance. Many nations forbid the importation of finished pharmaceuticals, and almost all require prior authorization of any medicinal import. Further, safety regulation may be manipulated to favor domestic producers.

A recent OECD report provides succinct characterization of the pharmaceutical trade policies of France:

> French policy offers an interesting example of the imaginative use of import restrictions. All pharmaceutical imports must be assembled and packaged in France. Ethical drugs require a visa from the French authorities before they can be marketed. To obtain a visa, the manufacturer must submit complete details of the production process and analytical control methods, together with the testimony of experts, drawn from a list of approved experts, concerning the safety and efficacy of the product. In practice, a visa is only granted if the material is produced and clinically tested in France. The visa system leaves much to the judgment of the individual examiner; it can be and apparently often is, applied in such a way as to favor French firms rather than the affiliates of foreign companies.[5]

An additional example of non-tariff barriers derives from price regulation for ethical drugs. In the United Kingdom, allowable prices have been based on the costs of bringing a drug to market, and in the mid-1970s, British authorities allowed research and development expenditures in the United Kingdom, but not elsewhere, to count as "costs" for the determination of price. Firms therefore had clear incentive to perform R&D in the United Kingdom that might otherwise have been executed in the United States. Continental price regulations have on several occasions been used to pressure U.S. firms into locating product facilities in western Europe.

An additional non-tariff barrier to U.S. exports of pharmaceuticals is, perversely enough, a U.S. policy. FDA regulations on new ethical drugs apply to exports as well as to domestic sales and, hence, prevent export of any new drug until it is approved for sale in the United States. This restriction holds even if the product has been formally approved for marketing in the importing nation. Given the substantial relative delay of the FDA in approving new drugs, transparent incentives exist for U.S. firms to manufacture new drugs abroad for sale, rather than export them from the U.S. production.

SUMMARY

This assessment of possible causes for the decline in U.S. pharmaceutical competitive position leads to several conclusions. In the first place, there are numerous similarities between the drop in pharmaceutical competitiveness and the general relative decline of the U.S. economy against Japan and western Europe. Specifically, deterioration in U.S. shares of pharmaceutical exports, national ethical drug sales, and some aspects of pharmaceutical innovation such as patents are matched by comparable relative declines in many U.S. industries, including others in the high-technology sector. Adverse shifts in these specific features of competitive position are thus best explained by the more vibrant multi-industry growth of foreign economies and not by factors specific to the ethical drug industry.

A second and quite important conclusion, however, is that two aspects of pharmaceutical competitive position have not followed this general trend but in fact have declined uniquely more severely. The U.S.-located share of worldwide production has dropped steadily throughout the 1960s and 1970s in a way that is unmatched by production shifts in other chemical industries. The explanation for the distinctive performance of pharmaceutical production is straightforward--more rapid growth of demand abroad coupled with widespread non-tariff barriers in other countries that effectively require domestic production and drastically reduce the viability of export strategies. The largest U.S.-owned firms have adapted to these developments by establishing production facilties abroad, and thus the decline in world-wide sales of U.S.-owned firms has been much less severe than the drop in U.S.-located production. The outcome here is a clear loss of jobs and income for the territorial United States and a disadvantage of indeterminate and possibly minor significance for U.S.-owned multinational firms.

The other pharmaceutical trend that differs from the general relative decline in the American economy is both more significant and of more ambiguous origins. The severity of the continuing drop in the world share of U.S.-located (and expectedly U.S.-owned) expenditures for pharmaceutical research and development is apparently unmatched in other related industries of our economy. This trend is particularly disturbing because R&D for new products is the foundation of the modern ethical drug industry and the essential basis for pharmaceutical competitive advantage. Traditional microeconomic factors such as labor costs or resource availability fail to explain this distinctive trend, and by process of elimination leads to government policies as the most likely cause--both U.S. government policies that relatively discourage pharmaceutical innovation and foreign policies that relatively encourage innovation abroad. To the extent that these public policies make the U.S. economy a less conducive environment for pharmaceutical innovation, all major ethical drug firms are affected because of the preeminent size of American consumption in the increasingly integrated global market for ethical drugs. U.S.-owned pharmaceutical firms are, however, affected more so because higher proportions of their sales and research are drawn from U.S.-located activities. In the end, the greater susceptibility of a corporation to U.S. government policies is the essence of what it means to be an "American" firm.

While the divergent government policies that have combined to make the United States a less hospitable environment for pharmaceutical innovation can be listed, it has not been possible within the limits of this study to determine the relative significance of each specific policy. Given the complexity of the issue, such detailed policy evaluation may never be feasible. This ambiguity should be a source of caution, but not of indecision in consideration of policy reforms. The economic stakes are large, and the issues raised are often quite general. While refusal to confront these issues is itself a policy option, the merits of such an option are dubious.

NOTES

1. David Weimer, "The Regulation of Therepeutic Drugs by the FDA: History, Criticisms, and Alternatives," Discussion Paper No. 8007, Public Policy Analysis Program, University of Rochester.

2. Alexander Schmidt, "The FDA Today: Cities, Congress, and Consumerism," speech to the National Press Club, Washington, D.C., October 29, 1974, cited in Henry Grabowski "Public Policy and Innovation: The Case of Pharmaceuticals," Technovation, 1982, pp. 157-189.

3. U.S. General Accounting Office, FDA Drug Approval--A Lengthy Process that Delays the Availability of Important New Drugs, HRD-80-64, May 1980.

4. Leonard Schriffin, "Lessons from the Drug Lag," report to Office of Technology Assessment, June 1980.

5. Organization for Economic Cooperation and Development, Impact of Multinational Enterprises on National Scientific Capacity: Pharmaceutical Industry, Paris, 1977.

4
New Developments Affecting the Industry

The industry structure and competitive performance in the pharmaceutical industry have not undergone radical change since the therapeutic revolution that occurred around World War II. While there is little prospect for developments of similar magnitude in the next decade or so, the future of the industry should not be regarded as completely predictable by simple extrapolation of past trends. Two features of the industry that promise to alter these trends arise from recent advances in biomedical sciences and from the growing innovational significance of Japanese-owned firms. A discussion of these follows.

SCIENTIFIC ADVANCES

Continued pharmaceutical innovation requires persistent expansion in the underlying scientific base. This expansion has so dramatically occurred in medicine, pharmacology, and chemistry that, according to numerous reports, the ethical drug industry verges on a burst of significant new products. Whether this development is imminent, knowledge in basic biomedical science has expanded so rapidly and consistently in the past few decades that steady progress in the knowledge required for drug development can be confidently predicted.

Recent upturns in the number of NCEs approved by the FDA for marketing in the United States have fueled this optimism (see Table 2.4, comparing the years around 1970 with more recent years). More important than this upturn (which in fact has been relatively modest and unlikely to be repeated) has been the spectacular success of a single new drug, Tagamet, introduced five years ago by Smith Kline and now the best-selling prescription drug in the United States (see Table 2.9). Apart from its tremendous financial success, the most interesting features of Tagamet concern the manner of its development, which is indicative of the changes in industrial pharmaceutical R&D generated by recent scientific advances.

Traditional pharmaceutical research depended in part on extensive screening of drugs with inevitable importance for serendipity in discovery of new medicines. While the popular misconception that this screening was conducted with little or no guidance from chemistry or biology is simply false, it is nonetheless true that the extent and quality of direction provided by basic science for pharmaceutical research have vastly improved. Gerald Laubach, President of Pfizer, Inc., summarized the impacts of improved scientific direction as follows:

> There is literally no comparison between the concepts and methodology that I had as a scientist in the 1950s and those that the present-day scientist brings to the task. Everything can be done more powerfully, efficiently, and incisively and that has made a difference in the qualitative potency of drug research and in the qualitative contribution of the products that are coming out.

The industrial effects of greater research precision may help a given dollar volume of research expenditures yield an increased number of INDs (or drugs entering clinical trials) and a given number of these INDs yield a greater number of actually marketed products. Prediction of future levels of marketed NCEs is thus complicated by the improved productivity of R&D, and simple extrapolations of past trends would provide underestimation of the volume of new drugs. Tagamet was designed using these new procedures, having been developed atom by atom to affect specific physiological processes.

If the cumulative industrial effects of improved pharmaceutical research increase productivity, this will serve to decrease the average cost of new drug discoveries and increase the earnings potential for these discoveries, thus indirectly encouraging additional research expenditures. From the perspective of this study, however, these positive developments, if they occur, will affect foreign-owned and located pharmaceutical firms largely to the same extent that they will comparable U.S. firms. Thus, while increased productivity may insure continued innovation and growth for the industry as a whole, it has essentially no implication for the relative position of U.S. firms.

Further, there are two important limitations to the impact of recent advances in basic science for the pharmaceutical industry. The first limitation arises in that the principal bottleneck in pharmaceutical R&D is not the generation of new substances with desired therapeutic effect, but rather the assurance that these substances are safe from adverse side effects. Advances in toxicology have lagged those in other biomedical sciences. Panel member Dr. Lewis Sarett, Senior Vice President of Merck & Co., Inc., addressed this issue.

I have been principally discussing new opportunities to achieve therapeutic specificity, ways to attack new diseases or to attack older ones with new weapons. The steady evolution of basic biomedical sciences over the past 50 years has made this task less and less difficult. The existence of many excellent medicinal chemical prototypes contributes heavily: penicillin, indomethacin, propranol and cimetidine spring immediately to mind. Where such prototypes are absent, new in vitro receptor methodology and clear perceptions of active sites of enzymes facilitate discovery of new prototypes. Improved instrumentation helps. But does this mean that we can expect a surge of new drugs? Not so, and in fact the statistics on new investigational drugs show that the number--at least of those originated in the U.S.--has declined even in quite recent years to a fraction of its earlier level. As you know, many reasons have been advanced for this: regulatory agencies, and escalating costs of development exacerbated by inflation over lengthy periods of time, for example. Undoubtedly these do contribute.

But I would submit that the rate-limiting stage of drug discovery has shifted away from efficacy--which, as I have said, is not so hard for the chemist and pharmacologist to achieve today--and toward safety, which is difficult to predict and even more difficult to control. Here I am not referring to safety in the gross sense. Of course, the record shows that new and experimental drugs have a superb history of safety when administered by the clinical pharmacologist under the guidelines of the animal toxicologist. But safety in the sense of freedom from the occasional potentially serious adverse reaction, a problem which does arise although infrequently in broader usage, is the frontier which now limits pharmaceutical innovation. Of course only a minority of new product candidates survive the preceding animal toxicology tests. Of the small group which does find its way into Phase I in the clinic, only 10 percent eventually survive to become marketed drugs. Is this for lack of efficacy? Not usually. I believe experience demonstrates that most fall by the wayside for reasons associated with unacceptable adverse reactions. Just from reading the newspapers, it is easy to recall examples of drugs which have started off bravely and faltered as toxicity began to manifest itself.[1]

Thus, both regulatory and product liability concerns about toxicity may impede new drug development.

A second limitation derives from the extensive supply of medicines. For many categories of disease, a pharmaceutical treatment of choice is already well established. Given the natural concerns over safety of new medicines for these categories, drug development shifts toward types of disease for which treatment is currently not efficacious. These research opportunities are, in general, more complicated and expensive than those addressed by past innovations.

JAPANESE DEVELOPMENTS

Traditionally, Japan has not been a significant presence in world pharmaceutical markets, largely because Japanese-owned firms were not at all successful at innovation (see Tables 2-1, 2-5, 2-6, 2-11, and 2-18). Several recent developments, however, have combined to increase dramatically the volume of Japanese drug discoveries and, thus, expected future sales. The first relevant development has been the large and continuing growth in research outlays by Japanese-owned pharmaceutical firms (see Tables 2-1 and 2-2). This growth has been encouraged by a major revision of Japanese patent law in 1975, extending coverage from "process" to "substance." Prior to this change, domestic Japanese firms could legally produce imitations of drugs sold by other firms so long as a unique process for production could be found, a relatively easy task. The new patent policy protects investments in research by preventing ready imitation. Additionally, pricing policies of the government-administered national health insurance system have been adjusted to systematically favor innovative products and to provide lower prices for more established drugs. This pricing structure provides continued incentives for research to discover new products. Finally, the pharmaceutical industry has been targeted by the government for international expansion, an action that lays groundwork for coordination of trade, pricing, and health-care policies to promote overseas expansion.[2]

The resulting rigor of the Japanese drug discovery effort may quickly dispel notions of the "imitative" or non-innovative nature of Japanese pharmaceutical firms. Actually, it is important to recognize that the imitative character of the late 1960s Japanese drug industry as a whole was the result of conscious policy decisions by public authorities. Faced with overwhelming foreign competition in this market, the logical first step for development of a competitive domestic industry was promotion of generic-type firms. This promotion was achieved by denying (mostly foreign) innovators adequate patent protection, by disadvantaging foreign firms through non-tariff trade barriers, and by generous pricing policies. Once a production-oriented domestic industry was flourishing, however, the Japanese government, in the mid-1970s, began to systematically skew a broad mix of policies, especially patent and pricing policies, in favor of more innovative firms. This second stage of Japanese industrial policy contrasts sharply with recent U.S actions, such as the eroding U.S. patent life (see Table 5-1) and progeneric pricing policies for cost-control.

The outcomes of these new corporate efforts at innovation and new government industrial policies is only now becoming visible in the world market. Yet those changes that are observable are quite dramatic.

- In 1981, Japanese firms ranked first in terms of the largest number of major new drugs introduced into world markets, being responsible for 17 out of 65 such products. U.S. firms, despite their larger aggregate research budgets, ranked second with 13 products launched. In third and fourth place were West German and Swiss firms, respectively, with 9 and 7 new products.[3]

- In 1982, Japanese firms accounted for over 16 percent of all U.S. patents issued for pharmaceutical and medicinal products, up substantially from previous levels (compare Table 2-6). Fully 38 percent of all U.S. medicinal patents granted to foreign firms in 1982 went to Japanese originators.[4]

- Japanese pharmaceutical houses are forming joint ventures with U.S. firms to market Japanese discoveries. Prominent among these arrangements are links between Abbott and Takeda, and Fujisama and Smith Kline. These linkages are clearly first steps toward direct marketing of Japanese products overseas.[5]

The long lags between strategic action by firms and governments and actual market impact in the pharmaceutical industry provide a substantial cushion for established American and European firms against any Japanese competition. But it must be remembered that if the Japanese targeting is successful, as it has been in so many other cases, then these same lags will make reversal of any Japanese gains exceptionally difficult.

NOTES

1. Lewis Sarrett, "Chemistry and Health," address to the AAAS Annual Meeting, January 1982.
2. Scrip, September 14, 1981.
3. Scrip, March 10, 1982.
4. Biotechnology Patent Digest, February 14, 1983.
5. "Innovative Japanese drugs move into the U.S.," May 10, 1982.

5
Options for American
Industrial Policy

The limited scope of this study has militated a focus on specifics of the pharmaceutical industry, away from the broader issues of industrial policy. Thus many critical questions are not addressed here, questions such as: what is the appropriate U.S. share of a major industry; what new policies must be pursued to attain this share; and how should the United States adjust its own policies in response to competitive policies of other nations? Questions of this sort are novel and complex and answers to them imply targets for overall American standards of living and portfolios of U.S. industrial strength. Some comments on these issues, however, are in order. On the one hand, some portion of the decline in U.S. pharmaceutical competitive position, particularly in the 1950s, was inevitable given postwar patterns of recovery and growth. Likewise, the probably 1990s emergence of significant multinational Japanese pharmaceutical competition with associated loss of U.S. market share may also be inevitable, representing a belated but natural entry of Japan in yet another high-technology industry. High levels of Japanese research spending and the size of their domestic market (second only to the United States) provide impetus to this development.

On the other hand, if the United States is to maintain standards of living comparable to those of other major industrial nations, it cannot suffer indefinite economy-wide declines in its share of world markets. It is now almost 40 years after World War II and more than a decade after certain western European nations equalled the United States in per capita national product. At some juncture, the continuous "natural" postwar relative decline of the U.S. economy must be regarded as "unnatural," and America must make efforts to maintain or strengthen its share of at least some industries. Given the historic economic position of U.S. firms in ethical drug markets, the great expense and long development time facing foreign entrants into the industry, the preeminence of American biomedical research, and the enormous U.S. government expenditures for health care, a continued decline of American pharmaceutical competitive position might well be regarded with concern.

Due to the limitations of this study, the policy options listed below have been chosen as much for their independent merit as for their advancement of the American position in pharmaceuticals. For example, the included proposals for FDA reform have been advanced elsewhere exclusively to improve delivery of medical care to American patients. The listed, limited proposals for trade and antitrust reform have been offered elsewhere for promotion of freer and more competitive markets. And patent reform has been suggested to restore equitable treatment of heavily regulated industries.

The policy options listed below are in no particular order.

TRADE OPTIONS

New ethical drugs are increasingly marketed on a worldwide basis, and the major U.S. pharmaceutical firms maintain extensive multinational operations. Under these circumstances, U.S. policies and responses to foreign policies on trade in ethical drugs have an important effect on development of the industry.

Due to a drafting error in the 1938 Federal Food Drug and Cosmetic Act, current U.S. law prohibits the export of new drugs (but not new antibiotics or biologicals) unless the FDA has previously approved these products for sale in the United States. This prohibition holds regardless of the comprehensive approval or not for marketing of a particular drug in the importing country by local regulatory authorities. This procedure provides an obvious incentive for firms of all nationalities to produce drugs outside the United States, particularly in light of the fact that FDA regulation tends to entail longer than average approval delays.

Nonetheless, the prohibition against exportation of unapproved new drugs applies only if the product is in fact labeled as a new drug. FDA could, through administrative action, permit the export of finished chemicals that are not specifically labeled for new drug use, in accordance with the export provisions of the 1938 Act. This would reverse the current disincentive to the production of these drugs in the United States and prevent the exportation of American pharmaceutical technology abroad:

• FDA should authorize the export of unlabeled chemicals for drug use abroad where a new drug application has not yet been approved in the United States.[1]

An additional barrier to trade arises in discriminatory foreign national policies. Simple retaliatory measures by the United States against these policies are likely to be counterproductive, yet the daily workings of foreign regulatory authorities are not an easy target of U.S. influence. Given the limitations of this study, only the following option may be offered:

• A detailed investigation should be undertaken as to the frequency and incidence of foreign pricing and regulatory practices that favor domestic products over American and other foreign origin products with provision for recommendations as to U.S. policy response.

DOMESTIC ECONOMIC OPTIONS

A diverse set of economic options are available for bolstering the competitive position of the U.S. pharmaceutical industry.

• The lengthy period of FDA regulatory review after granting of a patent but before marketing of the product eliminates a substantial proportion of the intended 17-year protection. As can be seen in Table 5-1, the average effective life of patented drugs currently approved for marketing has declined from 13.5 years in 1968 to 10.5 years in 1978, with a low of 6.8 years in 1981.[2] Patent protection for pharmaceutical innovation may be restored by extending the duration of each individual patent by the period of years consumed by the IND/NDA review. The restoration of patent time would simultaneously render investment in new drugs more attractive and expand the cash flow from which new drug research is needed. Patent restoration should be particularly important for "breakthrough" or therapeutically significant new drugs as these novel compounds often take longer than average for regulatory approval and hence currently receive an even shorter than average patent life.
• Traditional U.S. antitrust policy has discouraged mergers of small- and mid-size pharmaceutical firms in order to prevent industrial concentration. For example, the merger of U.S.-owned Parke-Davis and Warner Lambert was delayed for almost seven years due to antitrust concerns. Further, existing law does not stop the acquisition of small U.S.-located pharmaceutical firms by large foreign multinationals, instead it only hampers such acquisition by U.S.-owned firms. Reexamination of Table 2-15 demonstrates recent acquisitions by Hoechst, Bayer, and Ciba-Geigy, which were in 1980 the world's first, second, and third largest pharmaceutical firms, respectively. These firms further are

TABLE 5-1   Effective Patent Life for U.S.
Approved NCEs, 1966-1978

1966	13.6
1968	13.5
1970	14.4
1972	10.9
1974	13.0
1976	11.3
1978	10.5

NOTE: Average effective patent life is seventeen
years minus the average elapsed time for IND/NDA
approval.

SOURCE: Martin Eisman and William Wandell,
"Components of the Decline in Patent Protection
of New Drugs," *Research Management*, 1980.

themselves each subsidiaries of enormous chemical companies
with aggregate worldwide 1980 sales of $16.5 billion, $15.9 billion,
and $7.1 billion, respectively. Acquisitions are also listed by
SANOFI, a subsidiary of ELF Aquitane, the nationally owned
French petroleum firm (1980 sales worldwide of $18.4 billion). To
the extent that economies of scale in research have instead
shifted against smaller and even mid-size pharmaceutical firms, as
evidence seems in fact to suggest, then merger with either
domestic or foreign partners may be inevitable.

• A second option would be to allow research tax credit for
those research-related expenditures not now eligible for the
investment tax credit, allowing for appropriate carry-back and
carry-forward provisions. The current tax credit applies only to
investment in capital equipment.

• The government should also consider permitting research and
development expenditures incurred in the United States to be
allocated solely to the U.S. income of the taxpayer. Treasury
regulations recently issued to implement Sec 1.861-8 of the
Internal Revenue Code require that R&D expenditures be
apportioned to both foreign source and domestic income in an
effort to recognize the fact that innovations in the United States
often result in licensing and other revenue from foreign sources.
However the effects of the regulations are (1) to apportion
expenses to foreign source income even when that income is
incidental to the innovation; (2) to result in double taxation
because foreign governments do not allow this allocation to be
taken into account when figuring taxes due them; (3) to encourage
the location of R&D facilities abroad instead of at home to escape
the effects of the regulations, thereby diminishing both the
amount of R&D conducted in the United States and, in the

long term, the tax revenue generated from its conduct. Recent changes in tax law have provided temporary reprieve from these Treasury regulations. Consideration could be given to making the reprieve permanent.

• A formal study of product liability and its effects on U.S. innovation is needed. In the pharmaceutical industry, the financial threat of liability suits appears to have particularly severely affected those industry segments that develop and produce ethical drugs for healthy consumers (e.g., vaccines, antifertility agents, preventative medicines). Within the constraints of this study, it has not been possible to verify the impact of product liability or to identify appropriate policy responses.

• Current and emerging federal policies on the procurement and reimbursement of pharmaceutical products should provide fair recognition for innovation. Specifically, these government pricing policies should recognize research expenditures as a part of the cost of purchased pharmaceuticals.

REGULATORY OPTIONS

Regulation of new drug development in the United States directly affects the foundation of sales and earning growth in the pharmaceutical industry. This regulation is intended to permit the marketing of only safe and effective drugs and to prohibit the marketing of dangerous or ineffective drugs. Because of the enormous public pressure on FDA to be certain about its decisions, the agency requires a substantial amount of preclinical and clinical testing, closely scrutinizes new drug applications, and consumes substantial periods of time in the regulatory process.

Since 1955, it has been recognized that the new drug approval process needs revision. Dozens of investigations and analyses have been conducted, resulting in hundreds of specific conclusions and recommendations for improvement.

Most recently, at the request of Congress, an independent Commission on the Federal Drug Approval Process was convened to make specific recommendations for revising the current new drug review system. The report of that Commission contains several recommendations that would substantially improve the present process.

This study has not attempted to assess or quantify the public health impact of current FDA regulations, the impact of these regulations on the pharmaceutical industry, or the change that would be achieved by adopting the recent Commission recommendations. Even without such quantification, however, it is clear that the reforms of existing procedures and regulations recommended by the Commission would promote thorough, yet more rapid, review of new drugs.

All of these reforms could be adopted, in their entirety, through internal FDA reform by rewriting relevant regulations. Such a change in current regulations would be more expeditious than an attempt to adopt them through legislative enactment.

Thirteen of the more important Commission recommendations, and comments on the improvement they could make in pharmaceutical technology, follow.

The IND Process

Early Clinical Research

To permit more drugs to be tested more expeditiously, the preliminary requirements for introducing new chemical entities into humans (such as drug chemistry, animal toxicology, and protocol specificity) should be studied with the goal of simplifying them consistent with the protection of human safety.

Earlier introduction of new chemical entities into humans would reduce the time prior to NDA approval, thus reducing the required investment and increasing the industrial return.

Objectives of Investigational Drug Regulation

The IND regulatory system should recognize three distinct categories of regulated activity based on the purpose of each activity: (1) basic research, (2) drug development, and (3) therapeutic use of investigational drugs. In each category, the regulatory requirements should be rationalized and revised to meet the public policy objectives appropriate to the category.

By tailoring different regulatory requirements for each of these different types of IND submissions, FDA could focus its attention primarily on drugs intended for pharmaceutical development and thus expedite the new drug approval process.

Clinical Development Studies

The FDA should provide guidance to sponsors on the information needed from clinical development studies to support NDAs. On request, the FDA and outside experts should become more actively involved in the planning of clinical development trials.

Use of advisory committees and other outside experts to resolve
disputes on the type of evidence needed to prove safety and
effectiveness would, in particular, clarify FDA regulatory requirements and shorten the time needed to obtain an approved NDA.

Preclearance of Clinical Research

An experiment should be tried in which certain types of
clinical research, such as early clinical studies, are precleared
through either channel of a dual-channel regulatory system, in
which the sponsor of a drug may select either the FDA or a
non-government body that is subject to FDA standards and
monitoring to review the IND submission.

Use of expert panels outside FDA to review IND submissions for
clinical pharmacology research would encourage such research by
reducing the regulatory burden and cost.

The NDA Process

Application of the Standards for Drug Approval

The FDA Commissioner should clarify through regulation the
interpretation and application of the statutory and regulatory
standards for establishing substantial evidence of the effectiveness of a new drug. Effectiveness should be found to have
been demonstrated either by two--or when appropriate, one--
adequately designed and well-controlled studies that indivi-
dually provide the necessary substantial evidence of effectiveness or by the accumulation of data from adequately designed
and well-controlled studies that taken together provide such
substantial evidence.

By eliminating FDA's current rigid administrative requirement of
at least two adequate and well-controlled clinical studies,
approval of an NDA can often be expedited.

The NDA Submission and Its Review

NDA submissions should be greatly streamlined. Summary
presentations of data should replace the individual case-report
forms. The clinical sections of NDA submissions should be
designed to facilitate efficient review and should contain
tabulations of all the data needed to assess the design, con-

duct, and analysis of the important studies; they should not routinely include the individual case-report forms or tabulations of every datum those forms contain.

By reducing the bulk of the NDA submission, such submissions can be submitted earlier and reviewed more quickly, thus permitting a substantial reduction in the regulatory burden.

Utilization of Outside Expert Advice

The system of new drug development and approval should be modified to afford a more significant role to experts from outside the FDA, drawn from the academic and government biomedical research communities. On the request of the FDA or the drug sponsor, outside expert consultants should be available both to advise in the investigational development of each new drug entity or important new use of an approved drug and to participate in the review of its NDA.

As already noted, use of outside experts to resolve scientific policy issues will help assure better scientific decisions and expedite the entire NDA review process.

Application of the Standards for Drug Approval

The FDA Commissioner should revise the agency's review procedures, consistent with the Congressional intent, to insure that due weight be given in the approval process to the judgment of experts, including clinical investigators and medical specialists, as to whether the effectiveness standard has been met.

Data on drug efficacy and safety derived from studies conducted in other countries should be given full weight as evidence in proportion to their quality. The routine replication of randomized control trials should not be required; rather, when relevant foreign trials are convincing in themselves, it should be sufficient for a sponsor to provide clinical experience in the United States to supplement the results observed abroad.

These two recommendations would similarly reduce the need for duplicative clinical testing, which is extremely costly and time-consuming, and thus permit the earlier marketing of new drugs in this country.

FDA Management

Resources for the New Drug Review Process

Congress and the Administration should ensure that the FDA has sufficient and stable resources to review and act promptly on INDs and NDAs.

Adequate clerical personnel are particularly important to assure timely processing of documents and response to inquiries.

Improving Interactions with Industry

The FDA Commissioner should encourage the timely and equitable resolution of disputes regarding INDs and NDAs by ensuring that mechanisms appropriate for different types of disputes are in place and readily accessible. In particular, the Commissioner should expand on existing mechanisms and explicitly encourage their use. An ombudsman function should also be established in the Commissioner's Office where administrative or procedural problems can be taken when they cannot be satisfactorily or promptly resolved with line management.

An arms-length or adversary relationship between FDA and the pharmaceutical industry is not conducive to prompt and efficient handling of the NDA review process. Cooperation and a close working relationship must be fostered to expedite important regulatory decisions.

Tracking the Review Process to Ensure Timeliness

The FDA Commissioner should be provided with timely information on the status of NDAs undergoing review. Toward that end, a computer-based system for tracking NDAs should be utilized to enable the Office of the Commissioner, as well as senior Bureau of Drugs management, to review regularly the status of NDAs and to report appropriate information to the Commissioner.

Any efficient system for expediting review of NDAs must be based on an accurate tracking mechanism.

Conflict of Interest and Expert Advisers

Barriers to the participation of the best qualified outside expert advisers in the drug approval process must be addressed. In particular, the FDA Commissioner should request from the U.S. Department of Justice a less restrictive interpretation of the federal conflict-of-interest statute than that issued in 1978.

The use of outside experts to help assure sound scientific decisions and expedite the review process depends upon the availability of the best-qualified scientists in the field. Current restrictions unnecessarily limit the availability of many highly qualified experts whose advice could be important in the prompt resolution of important questions.

Improving Interactions with Industry

The FDA should provide guidance to its staff to encourage all review personnel to conduct timely, forthright, and even-handed discussions with sponsors of the significant issues that arise at any time during the IND and NDA review process. An atmosphere of mutual respect and professional conduct should be encouraged and maintained. Formal communication should be limited to situations where such formality is warranted.

By reducing formality and promoting greater interchange between industry and FDA, regulatory decisions can be discussed more freely and thus be made on a more sound and timely basis.

Adoption of these Commission recommendations, and the other related recommendations not reproduced here, will not assure an optimum drug regulatory process. They will, however, begin to address the serious deficiencies found in the present system without any loss of public health protection. To the extent that the system is improved, the pharmaceutical industry and therefore the public will gain immeasurably. Equally important, as incentives to invest in new pharmaceutical research are increased, greater gains can be expected in the discovery of new drugs that are effective in reducing the public burden of serious diseases that still remain to be conquered. Thus, the very economic incentives that will help return the U.S. pharmaceutical industry to its former stature will have important public health benefits as well.

NOTES

1. One member of the panel dissents from the panel support of this policy.

2. Statement of William M. Wardell to the Subcommittee on Investigations and Oversight of the Committee on Science and Technology, U.S. House of Representatives, February 4, 1982, p. 14.

3. For detailed assessment of the problems associated with the term of patents for drugs, see Statement of Peter Barton Hutt on Behalf of the Pharmaceutical Manufacturers' Association before the Subcommittee on Investigations and Oversight of the Committee on Science and Technology, U.S. House of Representatives, February 4, 1982; Office of Technology Assessment, Patent-Term Extension and the Pharmaceutical Industry, 1981; General Accounting Office, FDA Drug Approval--A Lengthy Process that Delays the Availability of Important New Drugs, Rep. No. HRD-80-64, 1980; Pracon, Inc., The Effective Patent Life of Pharmaceutical Products: Trends and Implications, 1978; and Eisman and Wardell, "The Decline in Effective Patent Life of New Drugs," Research Management, January 1981.

KENT BLAIR is a Vice-President of Donaldson, Lufkin & Jenrette Securities Corporation. He holds a B.A. from Colgate University and an M.S. in industrial administration from Carnegie-Mellon University. Mr. Kent was employed by Merck & Co., Inc. from 1963 to 1968, where he held a number of positions in sales, marketing, production, and finance. In 1968 he joined the institutional investment research firm of Auerbach, Pollack & Richardson as a Vice-President and Director, serving as the health industries analyst.

CHARLES C. EDWARDS is President of Scripps Clinic and Research Foundation. Dr. Edwards holds a B.A. and M.D. from the University of Colorado and an M.A. from the University of Minnesota. During his career, he has been in private practice as a surgeon, served on the staff of The George Washington University, and was a Vice-President and Managing Officer at Booz, Allen, & Hamilton. Dr. Edwards was also the Commissioner of the Food and Drug Administration from 1969 to 1973 and Assistant Secretary for Health, Education, and Welfare from 1973 to 1975. Dr. Edwards serves on a number of professional committees and boards including service as president of the National Health Council, trustee-at-large of the National Kidney Foundation, and a member of the Policy Advisory Committee of the National Board of Medical Examiners.

WILLIAM N. HUBBARD, JR. is President of the Upjohn Company, Inc. He received an A.B. from Columbia University, did postgraduate work at the University of North Carolina School of Medicine, and received an M.D. from New York University. Dr. Hubbard has taught at several universities and was the dean of the University of Michigan Medical School. He joined Upjohn in 1970 and has served as its President since 1974. Dr. Hubbard's professional memberships include the National Science Board of the National Science Foundation, the New York Academy of Medicine, and the Board of Regents of the National Library of Medicine.

PETER B. HUTT is a Partner in Covington and Burling. He received a B.A. from Yale, an L.L.B from Harvard, and an L.L.M. from New York University. From 1971 to 1975 he served as Chief Counsel for the FDA and Assistant General Counsel for HEW. Mr. Hutt is a member of a number of professional and honorary societies including the National Academy of Sciences' Institute of Medicine. He has received a number of awards including the HEW Distinguished Service award and the Underwood-Prescott award from MIT. He is the co-author of Dealing with Drug Abuse.

PHILIP R. LEE is Professor of Social Medicine and the Director of the Health Policy Program at the University of California Medical School, San Francisco. He received an A.B. from Stanford and an M.D. and M.S. from the University of Minnesota. After working several years at the New York University School of Medicine and the Palo Alto Medical Clinic, Dr. Lee served as Deputy Assistant Secretary and as Assistant Secretary of Health and Science Affairs at the U.S. Department of Health, Education, and Welfare from 1965 to 1969. He has received a number of awards including the AID Superior Honor award, is a member of several professional societies, and has contributed to a number of medical and scientific journals.

ARTHUR M. SACKLER is a research professor of Psychiatry at New York Medical College and the Publisher of the Medical Tribune Newspapers. He holds a B.S. and an M.D. from New York University. Dr. Sackler has been the founder or co-founder of a number of institutions including the Creedmoor Institute for Psychobiological Studies and the Sackler School of Medicine in Tel-Aviv. He has spent much of his career in psychiatry and in promoting medical communication as Chairman of the Board of the Medical Press, Inc. and President of the Physicians News Service, Inc. and the Medical Radio and TV Institute, Inc. Dr. Sackler has received a number of awards and is a member of numerous professional organizations including a Fellow of the American Psychiatric Association and an Associate of the Linus Pauling Institute of Science and Medicine.

LEWIS H. SARETT has recently retired as Senior Vice-President of Merck & Co., Inc. He holds a B.S. and D.Sc. from Northwestern and a Ph.D. from Princeton. Dr. Sarett has spent most of his professional career with Merck & Co., having worked as Assistant Director of Organic and Biological Research, as

Vice-President for Basic Research, and as President of Merck Sharp & Dohme Research Laboratories. A member of a number of professional and honorary organizations including the National Academy of Sciences, Dr. Sarett has received numerous awards including the Perkin Medal, the Merck Directors award, the William Scheele Lecture award of the Royal Pharmaceutical Institute in Sweden, the Northwestern Alumni Association Award of Merit, and an American Chemical Society award for creative work in synthetic organic chemistry. He has served on the Chemical and Engineering News editorial board and has contributed numerous articles to professional publications.

LACY GLENN THOMAS is Professor of Business at the Columbia University Graduate School of Business. He holds a B.A. from Vanderbilt University and an M.A. and Ph.D. from Duke University. Dr. Thomas held a position at the University of Illinois as Professor of Economics and was a research fellow at The Brookings Institution. He is the author or co-author of several papers and articles including "Estimating the Effects of Regulation on Innovation: An International Comparative Analysis of the Pharmaceutical Industry" published in the Journal of Law and Economics.

WILLIAM M. WARDELL is Professor of Pharmacology and Director of the Center for the Study of Drug Development at the University of Rochester Medical Center. He has attended Canterbury University and Otago University Medical School in New Zealand, holds a B.S. in physiology; a D.Phil. (Ph.D.) and a B.M., B.Ch. (M.D.) from the University of Oxford (U.K.); and a D.M. from the University of Rochester/Oxford. Dr. Wardell is a member of a number of professional and honorary societies including the Chairman of the Committee on Drug Development, National Council on Drugs, and the Committee on Certification in Clinical Pharmacology, American Society for Clinical Pharmacology and Therapeutics. He is the author or co-author of nearly 100 papers and articles and has received several awards and honors including the University of Oxford Radcliffe Prize for Research in Medical Science and the Merck International Fellowship in Clinical Pharmacology.

PAUL F. WEHRLE is Hastings Professor of Pediatrics at the University of Southern California and Director of Pediatrics, Pediatric Pavilion, Los Angeles County-University of Southern California Medical Center. He holds a B.S. from the University of Arizona and an M.D. from Tulane University. Dr. Wherle has held university appointments at the University of Pittsburgh, The Johns Hopkins University, and the State University of New York at Syracuse. He is author or co-author of over 190 contributions concerning infectious disease or environmental hazards and is a member of a number of professional societies including the Infectious Disease Society of America, the Society for Pediatric Research, the American Association for the Advancement of Science, and the American Academy of Pediatrics. Dr. Wehrle has served as a member of numerous committees and organizations including the World Health Organization and the Expert Committee on Viral Infections.

ALBERT P. WILLIAMS is Director of the Health Sciences Program at The Rand Corporation. His formal education includes a B.S. in engineering from the U.S. Naval Academy and an M.A. in international relations, an M.A. in economics, and a Ph.D. in economics from the Fletcher School of Tufts University. Before joining The Rand Corporation, Dr. Williams served on the White House staff and at the Bureau of the Budget. He received the Executive Office of the President Bureau of the Budget Professional Achievement Award and is a member of several professional organizations including the American Association for the Advancement of Science and the American Economic Association.

RICHARD D. WOOD is Chairman of the Board and Chief Executive Officer of Eli Lilly & Co. He holds a number of degrees including a B.S. from Purdue, an L.L.D. and M.B.A. from the University of Pennsylvania, an L.L.D from De Pauw University, an L.L.D. from the Philadelphia College of Pharmacy and Science, and a D.Sc. from Butler University. Dr. Wood has managed and directed Eli Lilly operations in Argentina, Mexico, and Central America. In 1970 he became President of Eli Lilly International Corporation and Chairman of the Board in 1973. Dr. Wood is a member of several business and professional organizations including the Business Roundtable, the Conference Board, and the Pharmaceutical Manufacturers' Association.

ALEJANDRO ZAFFARONI is President and Director of Research of ALZA Corporation. He holds a B.S. from the University of Montevideo and a Ph.D. in biochemistry from the University of Rochester. Before founding ALZA in 1968, Dr. Zaffaroni progressed from Associate Director of Biological Research to President of Syntex Laboratories and Syntex Research. In 1972 Dr. Zaffaroni founded Dynapol, a company with a primary aim of developing safe food additives. He has received numerous awards including election as a member of the National Academy of Sciences' Institute of Medicine, Fellow of the American Academy of Arts and Sciences, and Fellow of the American Pharmaceutical Association.

Index

This report on the pharmaceutical industry is one of seven industry-specific studies (listed below) that were conducted by the Committee on Technology and International Economic and Trade Issues. Each study provides a brief history of the industry, assesses the dynamic changes that have been occurring or are anticipated, and offers a series of policy options and scenarios to describe alternative futures of the industry.

The Competitive Status of the U.S. Auto Industry, ISBN 0-309-03289-X, 1982, 203 pages, $13.95

The Competitive Status of the U.S. Machine Tool Industry, ISBN 0-309-03394-2, 1983, 78 pages, $5.95

The Competitive Status of the U.S. Fibers, Textiles, and Apparel Complex, ISBN 0-309-03395-0, 1983, 90 pages, $7.95

The Competitive Status of the U.S. Electronics Industry, ISBN 0-309-03397-7, approx. 110 pages, $8.95 (prepublication price), available Fall 1983

The Competitive Status of the U.S. Ferrous Metals Industry, ISBN 0-309-03398-5, approx. 135 pages, $9.95 (prepublication price), available Winter 1983

The Competitive Status of the U.S. Civil Aviation Manufacturing Industry, ISBN 0-309-03399-3, approx. 120 pages, $9.25 (prepublication price), available Winter 1983

Also of interest...

International Competition in Advanced Technology: Decisions for America ". . . should help mobilize Government support for the nation's slipping technological and international trade position. . . . Leonard Silk, the New York Times. A blue-ribbon panel created by the National Academy of Sciences takes a critical look at the state of U.S. leadership in technological innovation and trade. ISBN   0-309-03379-9, 1983, 69 pages, $9.50

Technology, Trade, and the U.S. Economy, ISBN 0-309-02761-6, 1978, 169 pages, $9.75

Quantity discounts are available; please inquire for prices.

All orders and inquiries should be addressed to:

Sales Department
National Academy Press
2101 Constitution Avenue, NW
Washington, DC 20418